# A Guide to Hazard Identification Methods

# A Guide to Hazard Identification Methods

## Second Edition

Frank Crawley

ELSEVIER

Elsevier
Radarweg 29, PO Box 211, 1000 AE Amsterdam, Netherlands
The Boulevard, Langford Lane, Kidlington, Oxford OX5 1GB, United Kingdom
50 Hampshire Street, 5th Floor, Cambridge, MA 02139, United States



**Library of Congress Cataloging-in-Publication Data**
A catalog record for this book is available from the Library of Congress

**British Library Cataloguing-in-Publication Data**
A catalogue record for this book is available from the British Library

ISBN: 978-0-12-819543-7

For information on all Elsevier publications
visit our website at https://www.elsevier.com/books-and-journals

*Publisher:* Susan Dennis
*Acquisitions Editor:* Kostas KI Marinakis
*Editorial Project Manager:* Andrea Dulberger
*Production Project Manager:* R. Vijay Bharath
*Cover Designer:* Victoria Pearson

Typeset by SPi Global, India

# Contents

# Foreword

I am delighted to provide a foreword to the latest edition of this important book. Since the last edition was published, I have become much more aware of how important the thinking, which lies behind the methodologies in here, really is and how much more we need to do to share this thinking beyond the disciplines of process and chemical engineering.

IChemE's hugely successful Hazards conferences have now become a regular feature around the world and in 2020 the UK will celebrate its 30th Hazards conference. That means we have a huge bank of knowledge, which has built up in our profession on how to identify and manage process risks in a systematic way. The tools and techniques have been used repeatedly by generations of chemical engineers, they have been improved upon and new approaches have also been added to the toolkit as result of innovative ideas and lessons we have learned. It is great to see this book being updated to reflect the current state of knowledge.

I am also pleased to see the author extending his reach to share the applicability of the methodologies beyond chemical process safety. In 2017–18 I conducted an independent review for the UK Government into Building Regulations and Fire Safety in high-rise buildings following the tragic fire at Grenfell Tower in West London, which claimed 72 lives. In the process of gathering evidence for this review, it very quickly became apparent to me that many of the habits, which are so ingrained in us as chemical engineers, are not so common in other engineering disciplines.

Whilst our focus has been on how to design and operate chemical processes safely, we have been developing methods, which can equally be applied to many other dynamic and complex systems—including multiple-occupancy high-rise buildings. System design and management of change throughout the lifecycle are just as important in these areas as in chemical processes and the consequences of failure to take that disciplined approach can be catastrophic.

We must ensure that chemical engineers today maintain their focus on and competence in process and system safety, but we must do more than that. We must pass that knowledge on to future generations of chemical engineers through books, conferences, and other means of communication and education if it is to truly be part of who we are and what we do to deliver inherently safer processes. We must also take our knowledge to other engineering disciplines and share it freely with them. We have an important job to do—and a huge opportunity to make a difference—in helping others to see the value and the benefits of looking at systems and processes through the lenses of hazard identification and risk reduction.

I hope this book is truly widely read.

Dame Judith Hackitt

# Acknowledgements

This edition builds upon and extends the first edition published in 2003, which received contributions and support from many individuals and companies, which were acknowledged therein. Once again I have received suggestions, reviews of my drafts and general assistance from many present and former colleagues. In particular, I wish to express thanks to Richard Gowland, Ken First, Brian Tyler.

However, the final selection of material and its presentation is the responsibility of the author alone.

The author wishes to thank the following for permission to copy photos and text:

- Pictures in Chapter 15 supplied by courtesy of DNV GL (Spadeadam Research & Testing).
- IChemE for the re-use of various texts from the First edition and drawings as noted.
- DS Scott for the permission to copy the Risk Cube.

## Disclaimer

The author and collaborators have written this guide in good faith and do not accept any legal responsibility for misuse or misinterpretation of the text.

It is a guide to use and not the definitive document. The responsibility for the selection and accuracy and detail of any study lies with the user or a third party who might carry out the work on their behalf. The methods should always be selected with due thought to the need and the final outcome of the study. The user or third party should ensure that they have received adequate training and experience in their use.

# Acronyms and abbreviations

| | |
|---|---|
| **Actual MPPD** | Actual Maximum Probable Property Damage |
| **AIChemE** | American Institute of Chemical Engineers |
| **ACOP** | Approved Code of Practice |
| **ALARP** | as low as reasonably practicable |
| **ACMH** | Advisory Committee on Major Hazards |
| **BI** | business interruption |
| **BoD** | Basis of Design |
| **CEI** | Dow Chemical Exposure Index |
| **CHA** | Chemical Hazard Analysis |
| **CCF** | Common Cause Failure |
| **CMF** | Common Mode Failure |
| **CIMAH** | Control of Industrial Major Hazards |
| **COMAH** | Control of Major Accident Hazards |
| **COSHH** | Control of Substances Hazardous to Health Regulations 2002, amendment 2003 |
| **DIERS** | Design Institute for Emergency Relief Systems—AIChE |
| **ERPG** | Emergency Response Planning Guideline |
| **ESD** | Emergency Shutdown (system) |
| **EPSC** | European Process Safety Centre |
| **ETA** | event tree analysis |
| **F&EI** | Dow Fire and Explosion Index |
| **FDT** | fractional dead time |
| **FEED** | Front-End Engineering Design |
| **FTA** | fault tree analysis |
| **FME(C)A** | failure mode and effect (criticality) analysis |
| **HASWA** | Health and Safety at Work Act 1974 |
| **HAZID** | hazard identification (method) |
| **HAZOP** | hazard and operability (study) |
| **HIPS** | High-Integrity Protection System |
| **HS** | hazard study (0–7 in Chapter 3) |
| **HSE** | Health and Safety Executive |
| **IChemE** | Institution of Chemical Engineers |
| **IPC** | Integrated Pollution Control |
| **IPL** | Independent Protection Layer |
| **IPPC** | Integrated Pollution and Prevention Control |
| **LFL** | lower flammable limit (sometimes also explosive limit) |
| **LOPA** | Layer of Protection Analysis |
| **MAPP** | Major Accident Prevention Policy |
| **MF** | Material Factor |

| | |
|---|---|
| **MoC** | Management of Change |
| **MPDO** | Maximum Probable Days Outage |
| **MPPD** | Maximum Probable Property Damage |
| **NISHH** | Notification of Installations Handling Hazardous Substances (NIHSS) Regulations 1982 |
| **NPSH** | Net Positive Suction Head |
| **PFD** | Probability of Failure on Demand (aka Probability of Failure to Danger) |
| **PHA** | process hazard analysis |
| **PHR** | process hazard review |
| **P&ID** | piping and instrumentation diagram |
| **PPE** | personal protection equipment |
| **PtW** | Permit to Work |
| **RR** | relative ranking |
| **RAST** | risk assessment screening tool |
| **RIDDOR** | Reporting of Injuries, Diseases and Dangerous Occurrences Regulations 2013 |
| **SEP** | Surface Emissive Power |
| **SHE** | Safety, Health & Environment |
| **SIL** | Safety Integrity Level |
| **SIS** | Safety Instrumented (interlock) System |
| **SMS** | safety management systems |
| **UFL** | upper flammable limit (sometimes also explosive limit) |
| **VCE** | vapour cloud explosion |

# SI units

| | |
|---|---|
| **M** | mega |
| **m** | milli |
| **p** | pico |
| **F** | Farads |
| **J** | Joules |
| **Pa** | Pascals |
| **W** | Watts |

# 1

# Regulatory framework

Chapter outline



## 1.1 Overview

Society is subject to rules and there is no difference for industry. This chapter gives a brief review of the rules or Regulations by which the Chemical Industry must operate and links these into the various chapters within this Monograph. It concentrates on the United Kingdom, working within the European Union system.

There are many accounts covering the rules of different non-European Union different Nations or States. These must be understood fully and followed. Therefore, each nation or state within a nation must use the rules appropriate to that nation or state.

## 1.2 Background

The development of chemical and petrochemical manufacturing in the post Second World War era has led to a global expansion of production and storage facilities. For example, the size of olefin plants between 1948 and 1968 was doubling every 7 or 8 years, while the site footprint had hardly changed. New processes such as the direct oxidation of ethylene with pure oxygen were developed and many novel processes, particularly in the pharmaceutical industry, were brought on-line. These facilities introduced new hazards into communities and neighbourhoods which required regulation.

In general the historic legal framework was reactive, not anticipative, although newer frameworks have tended to be more anticipative.

The first 'Safety Acts' go back over 150 years with the regulation of chemical process safety becoming more established since the 1970s. Various reasons exist for the promulgation of the different regulations with one of the main initiators of new regulations

A Guide to Hazard Identification Methods. https://doi.org/10.1016/B978-0-12-819543-7.00001-X

or changes in existing regulations being the occurrence of major accidents with serious consequences either for the safety of employees, the neighbourhood or the environment.

Within United Kingdom there are some differences of approach to regulation within the component countries of the United Kingdom as a result of ancient Acts passed many years ago [1, 2, 3, 4]. Thus in the United Kingdom there is Environmental Agency, whilst in Scotland there is the Scottish Environmental Protection Agency (SEPA).

It is self-evident that the regulations require to be policed requiring a Regulator or Competent Authority with means of enforcement supplied by overarching Acts.

## Background in United Kingdom

The regulation of major accident hazards within the United Kingdom has its roots within the report of the Robens' Committee in 1972 [5]. This identified that 'major hazards' associated with technology present a particular problem, and recommended comprehensive provisions to deal with toxic, explosive and flammable substances be adopted. To this end the creation of a major hazards branch within the new inspectorate (i.e. Health and Safety Executive) was recommended, as was the establishment of a Standing Advisory Committee on Major Hazards (ACMH). The ACMH being given the remit to identify types of installations which have the potential to pose a major hazard and to advice on measures of control. One of the recommendations of ACMH (ACMH III) was 'Land Use Planning'. However before the resulting regulation from the Robens' Report, the Health and Safety at Work Act [6] could take effect, one of the worst accidents in British chemical processing history, the explosion and fire at Flixborough, occurred on 1 June 1974.

The sequence of major Regulations within the United Kingdom under HASWA has been:

- 1978 The publication of the Hazardous Installations (Notification and Survey Regulations) [7] as a consultative document in line with the Health and Safety Commission's duty to consult under the terms of the HASAW Act.
- 1983 Notification of Installations Handling Hazardous Substances (NIHSS) Regulations 1982 [8].
- 1984 RIDDOR Regulations (Reporting of Injuries, Deaths and Dangerous Occurrences Regulations).
- 1984 Control of Industrial Major Accident Hazards (CIMAH) Regulations (1984) [9] implementing EEC Directive 82/501/EEC.

In many ways the United Kingdom has led the European Union in Process Safety and uses a structure at four levels:

The Act
The Regulation—or Statutory Instrument (empowered by the Act)
The Approved Code of Practice (what is expected)
Guidance Notes

### Evolution European Union

Following an accident at a chemical facility in Italy a cloud of ca. 10 tonnes of chemicals, including an estimated 1–2 kg TCDD (dioxane), was released over the neighbouring town, Seveso. The name of the town has been adopted for the major European Union regulations. Hence 'Seveso Directive 1' and two updates following further incidents. The concept of two tiers of application was kept in both Seveso II and the current Seveso III Directive.

Member States were required to ensure that the lower-tier establishments provide notification to the competent authorities, fulfil the general duty of safe operation to prevent major accidents and that the operator established a major accident prevention policy (MAPP). The MAPP was the basis for the management of safety at Seveso III establishments.

The dangerous substances in the Seveso II Directive were primarily covered in ten generic categories related to acute toxic, flammable, explosive or environmental hazard properties. In addition the Commission proposed a very much shorter list of named substances than included in the original Seveso Directive. These are generally substances, which have a very widespread use in large quantities, are substances of particular concern or are of particular economic importance where a very low threshold level would cause an extremely large number of sites to be covered by the Directive.

## 1.3 Features of Seveso III Directive

The Seveso III Directive [10] places requirements on the Member States of the European Union to set national regulations for operators and for the competent authorities which enforce the national legislation and implement the requirements of the Directive. The requirements may be described as follows:

**Scope and definitions**: Application of the Directive is to 'establishments' in which sufficient quantities of dangerous substances are present or may be present. This may involve adding the various components in a multiprocess site. Activities, which are excluded from the scope, are those covering military installations and facilities; hazards due to ionising radiation; the transportation of dangerous chemicals and their immediate activities, marshalling yards, docks, wharves; waste land fill sites (excluding tailings ponds and dams); mineral extraction activities; offshore mineral and hydrocarbon extraction.

**General requirements**: Operators are to be required to show that they have taken all measures necessary to prevent the occurrence of major accidents and to limit their effects to man and the environment. Further, an operator should be able to prove to the competent authorities, at any time, that these measures have been taken. (See Chapters 3–5, 7, 15, 16, 17 and 19.)

**Notification**: Operators are to be required to notify the competent authorities of their existence, the hazardous substances involved, the activity, the person in charge of the establishment. This information is to be provided prior to commencing the activity and with any significant change, including closure of the establishment (see the 'Safety Report' later).

**Major Accident Prevention Policy (MAPP):** The operator is to be required to provide a written document setting out the policy for the prevention of major accidents and to ensure that it is properly implemented and guaranteed a high level of protection for man and the environment by appropriate means, structures and management systems. These means can be reviewed or examined by the methods outlined in Chapters 3–5 and in particular Task Analysis (Chapter 6). Audits (Chapter 17) then gives guidance on examination of these for weaknesses or systematic drift in standard.

**Safety Management System (SMS):** The operator is required to establish a Safety Management System in accordance with appropriate Annex of the Directive. The SMS should be proportionate to the hazards and risks of the activities of the establishment and adapted to the complexity of the establishment's organisational structure. The Major Accident Prevention Policy is an integrated element of the SMS. This gives the SMS a structure similar in form to that of environmental management systems according to ISO 14001 or quality management systems according to ISO 9001. [This Monograph does not cover management systems explicitly, but they are part of the analysis in Hazard Studies (Chapter 3), HAZOP (Chapter 4), HAZID (Chapter 5), Task Analysis (Chapter 6), Risk Assessment (Chapter 15) and Vulnerability (Chapter 16)].

**Safety Report:** The operator of an upper-tier establishment is required to produce a Safety Report, which is a core documentation of technical, organisational and management measures to ensure the safe operation of the establishment. The Safety Report is to have the purposes of:

**(a)** demonstrating that a major-accident prevention policy and a safety management system for implementing it have been put into effect in accordance with the information set out in Annex III;
**(b)** demonstrating that major-accident hazards have been identified and that the necessary measures have been taken to prevent such accidents and/or to limit their consequences for man and the environment;
**(c)** demonstrating that adequate safety and reliability have been incorporated into the design, construction, operation and maintenance of any installation, storage facility, equipment and infrastructure connected with its operation which are linked to major-accident hazards inside the establishment;
**(d)** demonstrating that internal emergency plans have been drawn up and supplying information to enable the external plan to be drawn up in order to take the necessary measures in the event of a major accident;
**(e)** providing sufficient information to the competent authorities to enable decisions to be made in terms of the site of new activities or developments around existing establishments.

The minimum contents of the Safety Report are listed in Annex II of the Directive. It must however contain an up-to-date inventory of dangerous substances in the establishment.

The Safety Report is not a document, which is finalised in one go. It should be a 'living document'. This means that a regular review process should be established, which ensures that that descriptions within the Safety Report are kept up to date. The Directive requires that the Safety Report should be amended as necessary as a result of alterations (material or significant change) to the establishment as the result of:

**(a)** new technical knowledge following 'near misses' or major accidents.
**(b)** a regular review process, which should take place at least every five years. [See PHR (Chapter 19) and other references to repeating studies regularly such as HAZOP (Chapter 4)].
**(c)** a significant change in the process and inventory.

In the case of 'Top Tier Processes' the Safety Case should be resubmitted every 5 years. The fact that a competent authority has accepted a Safety Report does not absolve an operator of any responsibility for the safe operation of the establishment, nor is it a statement that the establishment cannot have a major accident.

The Safety Report is to be submitted to the Competent Authority and this, in turn, is to give its opinion on the report, within a reasonable time. Practice has shown that this may range from: 'acceptance of the report as submitted'; through 'acceptance with amendments'; to 'rejection of the report'. ('Accepted but not necessarily acceptable has been quoted by the Regulator'.)

**On-site and off-site emergency planning:** Emergency planning is divided between those activities to be carried out by the operator, on-site, and those activities which fall within the responsibilities of the emergency planning authorities, off-site. The measures should be co-ordinated, which requires that the operator provides the emergency planning authorities with the necessary information to carry out the external emergency planning. The plans must also be developed in consultation with those who may be potentially affected. Therefore the drawing up and testing of internal emergency plans must involve the employees, including long-term subcontractors. External emergency plans should be drawn up in consultation with the public. The minimum contents of the emergency plans are given in the appropriate Annex of the Directive. The use of Risk Assessment (Chapter 15) and Vulnerability (Chapter 16) will assist in the development of these plans.

**Information to the public:** The Directive requires that Member States ensure that information is provided to the public and held in a permanently accessible electronic form about every establishment. Minimum content is listed in the appropriate Annex of the Directive. There is no stipulation as to where the information is to be held; thus, it is possible for the information to be provided on a single national portal or on the internet sites of establishments or corporations. In addition for upper-tier establishments information on safety measures and the required behaviour in the event of a major accident is to be supplied to those who may be affected. That is, the information is to be supplied to all persons and to all establishments serving the public without them requesting it. The practical application of this requirement in the Member States varies from the information being supplied by the authorities to the information being supplied by the operator of the establishment or a group of establishments which are in close proximity. Whichever approach is chosen, there needs to be a close cooperation between the operators and the authorities. There needs to be a clear definition of the neighbourhood, which is to be encompassed by the information and also how the neighbourhood should be warned in the event of an emergency. It should be clear whether the operator or the authorities trigger the alarm signal and how the third parties should respond.

**Reporting of major accidents:** Member States are required to ensure that in the event of a major accident the operator notifies the authorities as soon as practicable and by the most appropriate means. The notification shall cover the circumstances of the accident, the dangerous substances involved, the necessary data for assessing the effects of the accident and the emergency measures taken. Further the operator shall inform the authorities of the measures that are envisaged to deal with the effects of the accident and to prevent the recurrence of the accident. The information shall be updated by the operator in the event that additional facts or information become available.

The competent authority shall be required to ensure that urgent, medium- or long-term measures are taken to deal with the accident. The authority shall collect the necessary information for a full analysis of the accident. It shall also ensure that the operator takes the necessary remedial measures to deal with the accident and make recommendations on future preventative measures.

**Inspection of establishments by public authorities:** The Member States are required to ensure that the competent authorities establish an inspection system to assess the technical, organisational and management specific aspects of the safety of the establishment. This involves assessing the relevant documentation as well as carrying out an on-site inspection of the establishment. The Directive requires that the inspections of upper-tier establishments are carried out annually unless the competent authority has established a programme of inspections based on a systematic appraisal of the major accident hazards of the establishment. Following the inspection, a report shall be prepared and followed up with the management of the establishment.

The Directive does not make specific demands on the intensity of the inspection by the competent authorities. Experience shows that large, complex establishments cannot be inspected in their entirety in one on-site inspection visit. In addition it should be recognised that an inspection does not increase in effectiveness with increasing length and that general impressions, targeted inspections of individual aspects together with spot checks and interviews lead to the best results. Key for a successful on-site visit is good preparation beforehand, together with a transparent documentation of those aspects, which have been assessed and the findings of the inspection (see Chapter 17). The results need to be followed up with the management of the establishment and should include a timetable for the implementation of the necessary measures (see Chapter 17).

## Land-use planning

**Requirements on member states to provide information to the Commission**: For the Commission to be able to enforce the Directive as well as to be able to follow its application and develop the regulations in an appropriate manner, it is dependent on information provided by the Member States. The Member States are required to provide information relating to the following aspects:

**(a)** the implementation of the Directive within the national legal systems and structures.
**(b)** a 3-year report of the implementation of the Directive according to a questionnaire, which is provided according to a procedure on the standardising and rationalising of reports.
**(c)** a catalogue of the establishments falling under the Directive within the Member State.
**(d)** a reasoned list of the exceptions from the provision of information within the Safety Report.
**(e)** information on those major accidents which have occurred in the Member State. This information is to be provided according to an agreed format and is made public, except for those data which are declared as confidential, in the online database 'eMARS'.

## Other areas

There are close similarities between United Kingdom and European Regulations in Canada, Australia and New Zealand.

**Other nations have their own rules, so it is essential that the rules of that nation (or State in USA) are followed to the full.**

The Offshore Oil and Gas industry has some different rules, including the security of the Safe Haven (Temporary Refuge) against fire and smoke ingress as well as the means of escape from the installation as well as the site of a fire. The 'total risk' must also include helicopter transits from the heliport to the installation as well as shipping collision and structural failure of the installation. These do not apply onshore.

The helicopter transit is potentially a significant element in the total risk to offshore employees.

# Appendix

Further material may be found at:

aicheme.org 'Regulations and Resources page'.
J. Besserman and R.A. Mentzer, Review of global safety regulations, J. Loss Prev. Proc. Ind. 50 (2017) 165–183.
http://www.hse.gov.uk/pubns/hse49.pdf.
https://www.hse.gov.uk/pubns/hsc13.pdf.

## References

[1] The Planning (Hazardous Substances) Regulations 2015, 2015. SI No. 627.

[2] The Town and Country Planning (Hazardous Substances) (Scotland) Regulations 2015, 2015. SSI No 181.

[3] The Planning (Hazardous Substances) (Wales) Regulations 2015, 2015. No. 1597 (W. 196).

[4] The Planning (Hazardous Substances) (No. 2) Regulations (Northern Ireland), 2015. SR of NI 2015 No. 344 Statutory Instruments (Regulations) Owned and Enforced by HSE.

[5] L. Robens, Safety and Health at Work, Report of the Committee 1970-72, HMSO Cmnd. 5034, 1972.

[6] Health and Safety at Work Act 1974.

[7] Hazardous Installations (Notification and Survey Regulations) Regulations 1978.

[8] Notification of Installations Handling Hazardous Substances (NIHSS) Regulations 1982.

[9] Control of Industrial Major Accident Hazards (CIMAH) Regulations 1984.

[10] Directive 2012/18/EU of the European Parliament and of the Council of 4 July 2012 on the control of major-accident hazards involving dangerous substances, amending and subsequently repealing Council Directive 96/82/EC, Off. J. Eur. Union 197 (2012) 1–37. http://data.europa.eu/eli/dir/2012/18/oj.

2

# A guide to Hazard Identification Methods

## Chapter outline



## 2.1 Hazard Identification

Hazard Identification is fundamental to the safe design and operation of any system, be it a process plant or any other facility. The tools used vary according to the situation but they are all rigorous, systematic and depend to a greater or lesser extent on the knowledge of team members that will usually contain some form of expertise.

This book provides a selection guide that attempts to introduce, describe and illustrate the most commonly used methods, and then to direct the reader to more detailed references where fuller guidance may be found on the specific method. It does not provide a full listing of all of the available method. The selection in the 1st edition was based on an analysis of the more common methods used by EPSC Members: new and more modern methods have been added in this edition. It should be noted that the author found that a general method might have some very small, industry-specific, features or changes and then be known by a different name. The descriptions given here for each method are generic and non-industry specific, so that the methods can be used by different industries with minor, case specific variations. In the descriptions it will be seen that a method may sometimes use an adapted part of another method. This too is to be expected as most methods will include the best parts of other, tried and tested methods. The user should review the methods and chose the one most appropriate to that task.

The methods selected by the EPSC members include some that are on the boundary between identification and assessment. These include:

- Fault Tree Analysis (FTA)
- Event Tree Analysis (ETA)
- Failure Modes Effects (and Criticality) Analysis (FME(C)A)

These methods can be used both analytically and quantitatively. In the first mode they can be used as an investigative tool, but it is in the second mode that they have most use.

A Guide to Hazard Identification Methods. https://doi.org/10.1016/B978-0-12-819543-7.00002-1

Each method is described under the following headings:

- Definition
- Description
- Resource requirements (both personnel and documentation)
- Timing (against the PHA time frame)
- Advantages, disadvantages and uncertainties
- References
- Illustrated example (if needed)

Hazard Identification should be regarded as a set of tools that are used at the appropriate time during a particular project development (Chapter 3). The timing of use is not always critical and, other than possibly for HAZOP study, there may be several points in a project where a method can profitably be used. Also it should be recognised that a 'project' could be a maintenance task, a modification or a large construction project. The identification methods can be considered for any of these, the only differences being in the depth, the effort and the recording. While certain methods have more application to the process industry, from where they evolved, the methods can be of value in any task in any Engineering or Science field. The methods should always be selected with due thought to the need and the final outcome of the study.

The appendices in this book provide three examples showing which of the methods and it is intended that they should be easily understood. Full, real examples would be much more complex and detailed and cannot be used in an introductory document such as this. They also give a more detailed illustration of the interplay of the various methods and show how they can, and often do, interact to produce the best results. Where worked examples are given in the method descriptions in the main text they are intended to help explain the detailed working of that method.

All of the illustrative examples given in this Monograph are deliberately relatively simple, requiring little or no specific prior knowledge and therefore should be easily understood. Full, real examples would be much more complex and detailed and cannot be used in an introductory document such as this.

Table 2.1 shows the usual timings for each of the selected methods based on the standard stages of a project. As described in Chapter 4, an additional two specialised stages, number 0 and 7, have been included to the original set of 6. In the table the most likely time of use for a method is shown by an X: a less likely time is shown by (X). Definitions of the abbreviations are given in the appropriate chapters.

Note that HAZOP is shown as a possibility at project stage 2, as there may be some desire to carry out a 'preliminary' HAZOP on a piping and instrumentation diagram (P&ID) as it is being developed.

No index is provided for this book as a standardised format has been used in the description of each of the selected methods and the parts of each description are fully detailed on the contend pages.

**Table 2.1** Hazard Identification Method and indicative project (life cycle) phase (order as contents).

| | Project life cycle stages, 0–7 (as identified in Chapter 3) | | | | | | | |
|---|---|---|---|---|---|---|---|---|
| **Method used** | **0** | **1** | **2** | **3** | **4** | **5** | **6** | **7** |
| Hazard Studies | X | X | X | X | X | X | X | X |
| HAZOP | | | (X) | X | | | X | |
| HAZID | X | X | X | | | | | X |
| Task Analysis | | | (X) | X | | | (X) | X |
| LOPA | | | X | X | | | (X) | |
| Relative Ranking | X | X | (X) | | | | | (X) |
| Risk Analysis Screening Tool (RAST) | X | X | (X) | | | | | X |
| Check Lists | X | X | X | X | X | X | X | X |
| What If? | X | X | X | | | | | X |
| FMECA | | (X) | X | X | | | (X) | |
| FTA | | X | X | X | | | (X) | X |
| ETA | | X | X | X | | | (X) | X |
| Audits | X | X | X | X | X | X | X | X |
| Bow Tie | | X | X | | X | X | | X |

There are now many ‘Hazard Identification Methods’. The oldest two are probably ‘Check lists’ and ‘What if?’ The next oldest is probably ‘Hazard and Operability Studies’ (now known as HAZOP), which was trial run in 1968 and evolved from Method Study. Since then new methods have evolved, some are variations on other, older, methods such that there is the possibility of confusion in the minds of potential users. This monograph is a selection of the more commonly used methods giving the strengths and weaknesses of each and how/where it might be used. Each has an illustrative example of the use in reality. It is only a guide and is not the definitive document on each method but should give sufficient information for Engineers and Scientists to be able to chose which one might be appropriate in a specific case and to be able to assess the detail and quality of any work carried out by a second party study. In line with the disclaimer it is the responsibility of any potential user to seek training and guidance from a trained practitioner before using any specific method.

The general hierarchy of risk-reducing measures is as follows:

- Prevention,
- Detection,
- Control,
- Mitigation,
- Emergency response.

It is not possible to write a monograph such as this so as to cover every topic so it has had to be selective and leans towards ‘detection—at design’. In the case of prevention, mitigation and emergency response some guidance is given under Risk Assessment (Chapter 15) and Vulnerability (Chapter 16).

There is one final tool, which is called Process Hazard Review (Chapter 19). This is carried out on an existing process and examines the history and experience of the operations staff. It is not used in the development of a project but is useful for an existing process.

**Task Analysis** may be carried out throughout a project. The areas shown are indicative. It may be used during the Studies 2 and 3 process to design the maintenance of the equipment and at Study 7 to facilitate demolition. It will be used during all phases of operation including shut down and start up.

**Audits** can be carried out on any phase of the development of the project, but they are probably more useful during the steady-state operational phase of the project examining the drift of Management Systems from recognised standards.

Note that **HAZOP** is shown as a possibility at stage two of the Safety Studies—Process Hazard Analysis, as there may be some desire to carry out a 'preliminary' HAZOP on a P&ID as it is being developed.

## Ground rules

It is a timely warning that more and more of the HAZOP/HAZID studies are carried out by Facilitators who are 'contracted'. Too often they are chosen for particular skills and enter the study totally cold. The scope, terms of reference and ground rules should be agreed with the client prior to the study.

There are clear ground rules before executing any study. These are:

The Scope should cover the boundaries of the study, include what is to be done and what is to be excluded. It should be that result of a dialogue between the client and the Facilitator and should where appropriate cover the 'matrix'. It should cover who should carry out the actions and whether they should be ranked in priority. The scope should also address the responsibilities for restudy particularly following a HAZOP where there may be material changes in the drawings.

The Terms of Reference should cover the delivery date and where appropriate the references points, for example in an Audit against what should the actions be measured.

Reporting and Worksheets It is important that there is a standard form of worksheets or reports. This will greatly assist in the auditing of any study where the auditor will know where and what to look for in the final report.

## Interlinking

Many Hazard Identification Methods cross-link with others or have evolved from another. This will be more obvious when the links are shown. This has been drawn out in the three examples in Appendices.

In addition it should be recognised that some Hazard Identification Methods are appropriate to a specific task. That means that the correct method should be used so as to optimise the use of resources for the maximum returns.

These topics are covered in more detail in the chapters indicated.

The Legal Structure (Chapter 1) is derived from historic events leading to Acts or Regulations and the need to carry out a detailed assessment of a situation.

Hazard Studies (Chapter 3), also known as Process Hazard Analysis, evolved in the late 1960s following an upset on a chemical plant resulting in the release of chlorine. It is the systematic analysis of the development of a project from inception to demolition. It has a structure, which can be adapted for use in Task Analysis, writing a procedure, or checking the development of a project. Within this could be the methodologies called HAZOP, HAZID and Audits.

HAZOP (Chapter 4) is a structured analysis of a P&ID. It relies on the correct selection of 'guidewords' and 'parameters'. The resultant matrix is very much based on past world wide experience. It evolved from Method Study. In some respects it has similarities to Failure Modes and Effects Analysis. Once again the actions must be ranked in order of priority.

HAZID (Chapter 5) is a structured analysis, which again relies on a detailed understanding of the systems being analysed and the process architecture. Originally it was called 'Three-Dimensional HAZOP' in the 1980s. In some ways it can be thought of as a systematically derived *check list* of scenarios. Once defined each has to be subject to Quantified Risk Assessment.

Task Analysis (Chapter 6) is the analysis of a routine or infrequent operation and has some of the elements of hazard studies within it such as sequence and priority. It is probably best thought of as a checklist method as it relies on 'experience'.

LOPA (Chapter 7) is a simplified form of Risk Assessment, which has pre-defined conditional probabilities and failure rate data. It provides for the examination of major events and gives clear indications of the need for additional protective measures.

Relative Ranking (Chapter 8) is based on pre-prepared tools, which have built into them a set of parameters derived from the past world experience. This allows an element of filtering of the options before the project has evolved too far.

The Risk Assessment Screening tool (RAST) (Chapter 9) has some links to LOPA (Chapter 7) and again it is based on an accumulation of world experience and data.

Check lists (Chapter 10) are an aide memoire and are again based on historic events. These can be used in the development of a procedure or task analysis. They also relate to a form of HAZID.

What if? (Chapter 11) has some elements of HAZOP, which is cause driven (what if there is no flow?), and HAZID, which is effect driven (what if there is a major release of material X here?)

The one tool where the outcome of a failure is assessed in less detail is probably Failure Modes and Effects Analysis (Chapter 12), which uses subjective assessment skills. In some respects it is similar to HAZOP but is applied to a piece of hardware.

Fault Tree Analysis (Chapter 13) uses clearly defined logic of either a frequency or probability, which involves addition and multiplication to define the combination of events to define the 'top event' or the event of concern. It can be 'top down' (start with the final event) or 'bottom up' (create the logic, which will create the final event). The rules are well defined and the input data can be derived from reliability studies.

Event tree Analysis (Chapter 14) uses only probability data which are multiplied. It is often used with Fault Trees.

Quantified Risk Assessment (Chapter 15) feeds off vulnerability and consequence models. It also uses methods such as Fault Tree and Event Tree analysis.

Vulnerability (Chapter 16) completes the link to Risk Assessment and defines the possible injury/damage scenarios

Audits (Chapter 17) are a structured analysis of the corporate health, be it financial and/or safety, and attempt to identify the systematic erosion in standards.

Bow Tie (Chapter 18) has some links into HAZOP as well as Fault and Event Trees. It is a more visible (and more understandable) representation of causation and mitigation than Fault Trees and LOPA. The 'threats' use a more detailed analysis of the potential causes and examines some of the human factors in more detail than Quantified Risk Assessment. It also has links to Audits.

The next (four) methods are free standing and evolved from the need to examine high-risk applications such a Nuclear Power:

Risk assessment is the one link, which is common to all methods. Having identified the hazard or potential hazard it must be ranked. Some risk-ranking methods are better than others. As a trivial example a full Quantified Risk Assessment (QRA; Chapter 15) is not justified for a trivial case or routine operation. The assessment methods are:

1. Quantitative where the risk is given both magnitude and frequency using tool such as *Fault* Tree and Event Tree analysis and phenomenology. The error bands are about a half order of magnitude.
2. Semi-quantitative where the risk both magnitude and frequency is derived from predetermined data sets. Typically this applies to Layer of Protection Analysis and Ranking Tools. The error bands can be up to an order of magnitude.
3. Qualitative where the risk is based on experience and judgement. The error bands are relatively low as the risks are low by definition. Typically this may apply to Task Analysis.

Quantitative Risk Assessment will inevitably take information from human and hardware vulnerability data (Chapter 16). This may be of a physical or a psychological nature for humans. Likewise for equipment the data will be established in tabular format. The phenomenology is based on experimentally verified models such as gas (and where necessary liquid) dispersion, thermal radiation and blast from explosions. Finally the assessment must have a set of agreed criteria by which the risk can be assessed.

# 3

# Hazard Studies

## Chapter outline



## 3.1 Introduction

The topic referred to here as Hazard Studies, using the original ICI intent (1), has several other names. Common ones include 'Hazard Studies 1–6', Process Safety Reviews, Project Hazard Studies and Project SHE Reviews. Confusingly, PHA has also been used as an abbreviation for Preliminary Hazard Analysis.

## 3.2 Definition

Hazard Studies are a systematic analysis of a project, both process and control, by selected teams of experienced personnel, at defined phases during its development, to ensure that the safety standards built into the project satisfy corporate and national standards as well as the project, corporate and national criteria.

A Guide to Hazard Identification Methods. https://doi.org/10.1016/B978-0-12-819543-7.00003-3

## 3.3 Description

The most common set of studies, carried out as appropriate during the project life cycle, are:

**(1)** conceptual design
**(2)** front-end engineering (FEED) design—process
**(3)** detailed engineering design
**(4)** construction design verification
**(5)** precommissioning safety review
**(6)** post-start-up

The numbers are those referred to in the traditional Safety Studies, as first developed by ICI in the early 1970s [1]. 'Hazard Studies' has been used in his text. It is described by Swann [2] and adapted by many other companies. The composition of the team and the documentation changes from study to study as the project develops and the detail becomes more fixed. Descriptions of studies 1–6 have been given in the literature [3] with one account in some depth although specifically for the pharmaceutical industry [4].

In recent years, it has been customary to add two extra studies. These, numbered to leave the traditional study numbers 1–6 unchanged, are:

(0) inherently safer/less harmful to the environment
(7) decommissioning and demolition

It is important to recognise that these studies form a continuum and, to a degree, flow into one another. Each looks back at the outcomes of the preceding study and will feed into the following study. The timing is important so that each analysis is in place for the correct phase of the project but is not carried out too early, when the definition may still be incomplete, or too late, when the options for change may be limited. The timing of stages 1–6 in a normal project life cycle is shown in Fig. 3.1. Obviously study 0 inherency precedes study 1 and demolition or abandonment follows study 6.

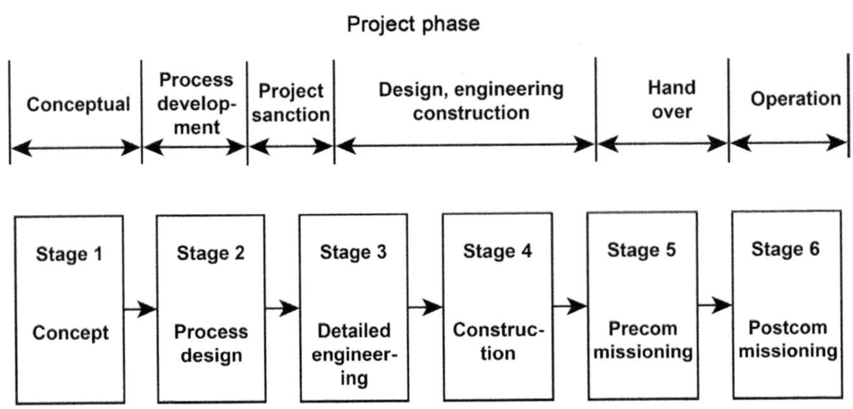


**FIG. 3.1** Sequence or hazard studies in a project.

It is important to realise that sanction will be granted in between studies 2 and 3. This value is usually ±10%. If significant issues are not identified before sanction, there is a grave risk of cost overrun and/or project delays.

## Study 0: Inherent safety

This study is carried out as the design options are being identified; its aim is to examine and to apply where possible the principles of inherently safe design [5]. The team will be small and the study will only last about 1 day. Members will include a project representative together with a process engineer, chemist, safety engineer and a technology expert. Ideally there would have been a careful review of the history of that type of process, including a search of accident databases.

The inherently safer process starts with guidewords, typically:

- Intensify
- Attenuate
- Simplify
- Get it right first time
- Change
- Substitute
- Eliminate

**Intensify**: Concentrate the process in a smaller, higher pressure reactor and reduce the working inventory or total leak potential. An example might be a high-pressure catalytic reactor, which is significantly smaller than the conventional low pressure reactor. Another might be the use of a linear reactor instead of a continuously stirred back mixed reactor. Another might be the use of 'green-field' equipment, which has by the very nature of the design a very low inventory; some of the modern compact heat exchangers would fit into this heading. The end point is that while the peak outflow from a hole (loss of containment) may be higher the total outflow will be significantly lower.

**Attenuate**: Reduce the working pressure/temperature such that the leak rate—should it occur—is less or less likely to ignite/explode. An example might be the use of refrigerated storage of cryogenics instead of 'green-field' storage. Once again the use of a catalyst lends to inherency.

**Simplify**: This is self-evident

**Get it right first time**: Avoid the need for last-minute change or even recognising the whole spectrum of conditions, which may apply to choosing the correct materials for fabrication and the choice of design pressure for equipment. It can also mean 'de-clutter' the process and avoid a surfeit of 'add-on safety features', which do little for SHE or efficiency but create operational problems.

**Change**: While the concept of change is simple, it does require a bit of thought! Consider the '***change***' in a layout such as to segregate flammable materials from sources of ignition or the positioning of a valve such that access is enhanced—the layout or access is then inherently safer. Change may involve a new process if the environmental implications were adverse. 'Change' is simple, but finding the solution is less so!

**Substitute**: Change the process route using chemicals, which are safer or which do not produce hazardous by-products or intermediates. Steam is inherently safer than hot oil. Steam heating may be inherently safer than electrical heating in that it has a self-limiting upper temperature.

**Eliminate**: This is more a statement of the obvious. Consider the design pressures; can you eliminate the need for overpressure protection by the selection of the equipment design pressures?

**Eliminate** and **Change** look at the same basics problem from different directions.

Other aspects to think about at this stage are the ability to recover from and to survive an upset or to tolerate the extremes of the operating/upset conditions envelope and, considering the environment, how to capture leakage and rework it.

## Study 1: Concept stage hazard review

In this study the basic SHE data for the operation should be reviewed and SHE criteria set. This study should identify what further information is needed, and define the program of studies required to ensure that all safety, health and environmental issues can be adequately addressed during the project. It should also check if there are any fundamental issues that might stop project development. Typical issues would include:

- properties of the feed stock, product and by-products, toxicology and handling,
- operating/start-up conditions,
- reaction kinetics/reactor control—exothermic or endothermic,
- special overpressure protection particularly at high/low pressure interfaces,
- inherently safer heating (cooling) systems,
- special construction considerations,
- containment integrity,
- environmental impact,
- site utilisation and appropriateness,
- source of materials and any transport implications,
- water availability,
- off-site services.

In addition, any constraints due to relevant legislation should be identified. A decision may be taken on which of the remaining Hazard Studies (2–6) should also be undertaken. Efforts to apply the principles of inherent safety should continue.

On occasion there may be important location-specific issues that must be identified. There would, for example, be some very different issues with a green-field, joint-venture site in a remote, developing country compared to those at an established, wholly owned site in a developed country.

The team required for this type of study would have a similar composition to that of study 0, but the work may extend over a number of days. The results of the study may lead to the project being delayed or even abandoned if there are major concerns about SHE issues.

## Study 2: FEED/process definition

This study is carried out as the design option is being developed to the point where detailed design is beginning. It will typically cover hazard identification and risk assessment, and will consider the operability and control features that must be built into the detailed design to manage these risks. The study should look for any residual concerns that might delay the design of the project. Typical issues would include:

- establishing the project safety criteria,
- carrying out the start-up/shutdown analysis,
- reviewing the control of exothermic reactor,
- reviewing the production of hazardous bi-products in the reactor—consider the temperature extremes,
- questioning whether new toxicological studies are required?
- carrying out the preliminary risk assessments (loss of containment) leading to the safety case,
- carrying out the over/under pressure analysis,
- carrying out the preliminary assessment of the performance of the shutdown system, for example by LOPA,
- considering the safety impact of existing systems on the new system and the impact of the new system on the existing systems,
- considering the siting of occupied buildings both on and off site,
- considering any special environmental aspects,
- regulatory approvals.

The study team would be similar to that for studies 0 and 1 but would also include operations personnel and specialists as appropriate. A block flow diagram should be used and the duration of this study could be a number of days.

The outcome of this study may lead to the project being delayed or even abandoned if there are concerns for SHE issues. However, this is less likely than after study 1. Once the project has reached this stage, the cost of any major remedial work in the future is likely to be high so there should be no major outstanding issues at the end of this study.

It is important that the major safety issues are addressed and designed out at this stage as the cost implications of adding more protective systems during the detailed design must not be underestimated.

It is during this stage that the first Risk Assessments will be carried out particularly if they are required for the production of a Safety Case.

## Study 3: Detailed engineering design

This study involves a detailed review of a firm design aimed at the identification of hazard and operability problems and questions whether there are any faults in the detailed design. HAZOP studies [6] are normally carried out at this stage. Typical aspects would include:

- confirmation that the design satisfies the project criteria,
- reviewing isolation standards for removal of equipment for maintenance*
- reviewing preparation requirements for removal of equipment for maintenance*
- materials-handling studies, including how equipment is removed for maintenance and it's access*
- HAZOP studies,
- evacuation studies,
- relief and blow down studies,
- hazardous area classification,
- personal protection and manual handling,
- starting the writing of the operating/start-up/shutdown instructions,
- final detailed risk assessments,

*Note*: *Must be carried out prior to the HAZOP.

The team members would include some from study 2 but might also include mechanical/instrument/electrical engineers. The duration is difficult to define, as there will be different demands at different times during the design. The most intensive will be the Hazard and Operability Studies and the Relief and Blow Down Study. This stage is the most intensive as the design is hardening and construction will be starting. Any changes after construction has started are likely to be expensive and time consuming.

## Study 4: Construction/design verification

This review is performed towards the end of the construction stage. The hardware is checked to ensure that it has been built as the designer intended and that there are no deviations from the design intent. It is also confirmed that the actions from hazard study number 3 have been incorporated and operating and emergency procedures are checked. Typical issues would include:

- as built drawings,
- 'punch/reservation lists',
- commissioning/start-up procedures,
- training and team selection.

The study is time consuming and its different parts are carried out by experienced individuals or by small teams, as appropriate. The commissioning team must be represented. Checklists may be used as aids during this study. This study flows directly and easily into study 5.

## Study 5: Precommissioning safety review

This study is carried out just before commissioning and checks if all construction issues have been finalised and whether the commissioning team is ready for start-up. It could be viewed as a form of Audit (Chapter 17). Typical issues would include:

- operating procedures,
- pre-start-up preparation (cleaning, drying, purging, equipment testing, shutdown system function testing),
- operator training/practices complete,
- all safety equipment in place,
- compliance with company and legislative standards,
- checking management of change procedures, including checks on changes done at earlier stages

This can be of similar length to study 4 and is carried out in the same manner. Company SHE specialists can be included in the team. It is the last opportunity to correct any deficiencies in the hardware and software.

## Study 6: Post-start-up review

This study is carried out 6–12 months after the start-up and asks 'what lessons can be learned, good or bad, from the project?' The team should include the commissioning manager, the plant manager and plant engineer, the project manager and other specialists as required. It should not last for more than a few days. It is important that this study is used as a basis of improving future project and design systems.

## Study 7: Decommissioning and demolition

This study has many of the features of the studies 1 and 2. It seeks to identify what could go wrong during the decommissioning and demolition phases. The team should include expertise in the process's design and operation as well as in demolition. The length of the study will depend upon the complexity and size of the plant, but it may take several days. Typical issues would include:

- cleaning,
- possible residual fluids and the associated risks and methods of re-use or disposal,
- removal and recycling of catalysts,
- making the plant safe for demolition, and briefing the demolition contractor,
- access for demolition, including any rusted ladders/walkways/loose gratings/rusted lifting beams, etc.,
- order of demolition,
- special services require for demolition,
- paperwork hand over from the decommissioning team including procedures for recording the process and recording any tests that are done,

- mode of size reduction and cleanliness requirements,
- maximising recycled materials, including power cables, stainless or high alloy steels,
- dumped materials,
- hazards associated with the 'waste materials' (asbestos/catalysts/materials caught in the drains),
- how to deal with foundations such as concrete,
- order of demolition especially of supports,
- carrying out a HAZID study of the overall process.

## 3.4 Resource requirements

### Manpower

The manpower will vary both in numbers but also in skills and hours during the various phases of a project. Indicative manpower and duration have been given above for each study. It may be appropriate to co-opt nonproject members into the study team to add specialised skills. In the case of study number 3 it may be appropriate for the several of the team to be selected from outside the project.

### Documentation

The documentation will vary according to the project development and the study itself. The initial studies may only have a project/process outline.

## 3.5 Timing

Indicative timing for each study has been given.

## 3.6 Advantages, disadvantages and uncertainties

**Advantages:**

- There is a clear documentation of the order and detail of the hazard identification process.
- Expertise can be brought into the appropriate team to supplement and support the project team effort.
- The recording of the studies will assist the preparation of any Safety Case.
- Properly coordinated and timed studies will avoid wasted effort, reduce costs and will ensure a better project safety standard.

**Disadvantages:**

- The main disadvantage is that the timing of the work has to be carefully assessed to ensure that there is the maximum of returns for the effort expended.

- There are some perceptions that the HAZOP, in particular, is a hindrance to the project development. This is untrue for a properly run study.
- There is a perception that the studies do not give a good reward for effort/money. This, too, is untrue for properly run studies.

**Uncertainties:**

- In summary there are no uncertainties if the timings are correct, the studies are well planned and suitable team members are used at each stage.

## 3.7 Applications

The studies indicated can be used on a project or, in a more simple form, they can be used to prepare a permit to work, an experimental program, a process modification or even a change in an operating parameter. The resources and duration will be different but the rigour and documentation will give assurance of due diligence.

## References

[1] ICI Mond Division Hazard and Reliability Training Course (Confidential to attendees and no longer available) This reference is given as it is the first recorded evidence of "Hazard Studies" or PHA.

[2] C.D. Swann, M.L. Preston, Twenty five years of HAZOPs, J. Loss Prev. Process Ind. 8 (Part 6) (1995) 349–353.

[3] F.P. Lees, Loss Prevention in the Process Industries, second ed., Butterworth-Heinemann, ISBN: 0-7506-1547-8, 1996.

[4] J.E. Gillett, Hazard Study and Risk Assessment in the Pharmaceutical Industry, Interpharm Press, Illinois, ISBN: 1-57491-029-9, 1997.

[5] T.A. Kletz, Process Plants—A Handbook for Inherently Safer Design, second ed., Taylor & Francis, Philadelphia, USA, 1998.

[6] F. Crawley, B. Tyler, HAZOP Guide to Best Practice, third ed., Elsevier, ISBN: 978-0-323-39460-4, 2015.

## Further reading

[7] A worked example is given in Appendices.

# 4

# Hazard and operability study (HAZOP)

Chapter outline



## 4.1 Definition

A HAZOP study is a structured analysis of a system, process or operation, carried out by a multidisciplinary team. The team proceeds on a line-by-line or stage-by-stage examination of a firm design for the process or operation. This is done by using a set of guidewords in combination with the system parameters to seek deviations from the design intention. Where a potential hazard or operability problem is found the team use their experience to decide whether changes or further investigations are required. A written report is prepared.

## 4.2 Description

HAZOP study is one of the most widely used hazard identification methods within the chemical (and many other) industries. The technique was developed in the late 1960s

**A Guide to Hazard Identification Methods. https://doi.org/10.1016/B978-0-12-819543-7.00004-5**

and became widely used after the publication in 1977 of the Chemical Industries Association guide [1]. Detailed descriptions and examples are also available in several texts [2, 3] and a recent monograph [4].

A study can only be done when a detailed process description and a full design is available and should only be done when these are in a well defined and frozen state. Indeed it is important that any subsequent changes are only those necessitated by the HAZOP study findings or are made under a strict management of change procedure. Study boundaries must be defined covering the items of equipment to be examined and the modes of operation that need to be examined. The remit should also make clear, which types of potential problem are to be sought. The actual study is done by a team of experienced staff, mostly from the plant itself, chosen for their knowledge of the plant, the process and site, the operating and control systems, the potential hazards and other problems. The process or operation is divided into sections or stages and which are examined in turn by the team. A design intention is developed for the section and then possible deviations, such as more flow or higher temperature (Table 4.1) are considered. To assist in the identification of possible deviations a set of guidewords are used. Examples are *more of, as well as* and *reverse* (Table 4.2). A deviation is obtained by linking a guideword to a parameter of the system, such as flow, temperature or composition. If a cause can be suggested for that deviation then

**Table 4.1** Examples of some possible parameters for process operations.

| | | | | |
|---|---|---|---|---|
| Flow | Pressure | Temperature | Mixing | Stirring |
| Transfer | Level | Viscosity | Reaction | Composition |
| Addition | Separation | Time | Phase | Speed |
| Particle size | Measure | Control | pH | Sequence |
| Communication | Start/stop | Operate | Maintain | Services |

**Table 4.2** Standard guidewords and their generic meanings.

| Guideword | Meaning |
|---|---|
| No (not, none) | None of the design intent is achieved |
| More (more of, higher) | Quantitative increase in a parameter |
| Less (less of, lower) | Quantitative decrease in a parameter |
| As well as (more than) | An additional activity occurs |
| Part of | Only some of the design intention is achieved |
| Reverse | Logical opposite of the design intention occurs |
| Other than (other) | Complete substitution. Another activity takes place |
| Other useful guidewords include | |
| Where else | Applicable for flows, transfers, sources and destinations |
| Before/after | The step (or some part of it) is effected out of sequence |
| Early/late | The timing is different from the intention |
| Faster/slower | The step is done/not done with the right timing |

**Table 4.3** Suggested worksheet layout for the standard recording worksheet.

Sample worksheet layout:

Date:
P&ID number (and update number):
Attendees:
Intent of first line:

| Ref no. | Parameter | Guideword | Cause | Effect | Protective systems | Action | On |
|---|---|---|---|---|---|---|---|
| | | | | | | | |

the effects or consequences are analysed and the team must decide whether the existing protective systems or safeguards reduce the risk to an acceptable level. If this is not the case, or if there is any uncertainty about a part of the analysis, a recommendation for action is recorded with the reference number, parameter, guideword, cause and protective systems as shown in the worksheet Table 4.3. Responses to the recommendations are reported back to the team and incorporated in the final report.

Important elements in a HAZOP study include its specification, the team make-up, the preparation for the study, the detailed analysis during team meetings and the preparation of the report. These are considered below.

## Specification

The specification of a study should set out a clear remit for the team. A clear indication is needed of the physical boundaries of the plant to be included and of the modes of operation that should be considered, the Scope. These might include start-up and shutdown for a continuous operation and batch operations such as catalyst regeneration. For a batch process the special situations might be manual control of a stage or reworking of substandard product. The time taken by the study is greatly affected by the number of conditions that must be covered. The types of problem to be sought and reported should be made clear. All studies will search for potential hazards and most will also seek out operability problems. The definition of hazards may include environmental problems and operability, may include reliability and any deviation that could cause a financial loss. It is also important that the relationship to the project management team is spelt out so it is clear who initiated the study, to whom the team reports and who has responsibility for chasing-up responses and ensuring recommendations are followed through.

## The team

The composition and quality of the team is one of the most important factors in determining the outcome of the study. Team sizes vary greatly. A small, short study may require only four members. For major studies a team of 6–10, including a leader and a scribe, would be more common. If more than 10 are identified, the reasons for such a large number should be scrutinised since it means the meetings are very costly and are more difficult to control and run efficiently as two discussions can run in parallel.

All studies need a designated leader (often called facilitator) and a secretary (often called scribe). Tasks for the team leader include checking, and perhaps helping with, the preparation of material for the study, ensuring all team members are briefed and understand how the study will be conducted, leading the team in the application of the HAZOP methodology and helping guide the team into a thorough analysis of the problems that are identified. A good team leader will be experienced in HAZOP studies and have good human relationship skills. It helps if the leader is independent of the project and can rely upon the members to do the bulk of the technical analysis. Obviously the team leader must have sufficient technical knowledge to understand the discussion and to be able to judge when it becomes either superficial or trivial. Team leaders should have a wide experience of HAZOP studies and will benefit from formal training.

The team members should be chosen for a combination of their detailed knowledge of the process or operation being studied, their experience of similar process or operations and the background expertise they can bring to the discussions. Individuals might be selected from the design group, operations and maintenance, control and instrumentation, the safety group or to bring specialist knowledge about a particular item of equipment or aspect of the operation. Between them they should have many years of relevant experience and come from a variety of backgrounds. These might include chemical or process engineering, chemistry, mechanical engineering, electrical engineering or computer systems. Some companies include an independent process engineer in the team to aid the critical examination of the design. Each study requires a particular combination to give confidence that identified problems can be competently analysed and that reasonable judgements will be made. All team members should either be experienced in the HAZOP study method or have received adequate training. Most team members will be present at all of the meetings to ensure continuity but specialists, such as a piping engineer, may be included for part or all of a meeting where detailed knowledge is required.

## The leader/facilitator

The following are some useful rules for the facilitator:

- Always maintain eye contact.
- Watch for the body language of the team as even the odd raised eyebrow or the shake of a head means something. Someone is not in agreement and the reason for this must be explored. They may be right but are unwilling to challenge a senior person! Maybe they are shy or unsure of their role. Draw them into the discussion and they will grow in confidence and contribution.
- Look out for the 'quiet' member and draw him/her into the discussion. (See tools for eliciting information.)
- Watch your own body language. Are you aggressive or passive? Do you have a 'friendly approach'? There will be a point in the discussion where the team must reach a conclusion. This must be handled with sensitivity.

- Never identify an issue before the team has found it. In some processes there are only a limited number of issues but in other process industries there may be many and as well as obscure ones. The leader's role is to 'facilitate' the team to find the issue using prompts where necessary. A good facilitator, with a depth of experience, will probably have seen the issue previously in one form or other so he/she has to 'tease out' the answer. (Obviously if the facilitator finds all of the problems the team will feel demotivated and not participate in the discussion.)
- See Audits (Chapter 17) for clues as to facilitation.
- Sometimes it might be useful to use the HAZID approach to flush out an issue. It also keeps the team on its toes!

## Preparation

It is important that the parameters (Table 4.1) and guidewords (Table 4.2) are drawn up in a case specific manner. These should then be put into a matrix which describes the realistic combinations of the two and those combinations which are physically impossible. This will eliminate wasted time when a particular combination is discussed, and the final conclusion is that combination cannot exist. Table 4.4 shows an illustrative example for a simple house water system.

In addition to definition of the study remit (Scope) and team selection it is essential that adequate documentation is ready and available. The team needs to be able to visualise the plant in its completed state and to know in detail how it will be operated and controlled. Material properties, reactions and conditions must be known as well as the obvious hazards associated with the operation. Information from earlier hazard studies should be available as well as the design specification. Some of this will be readily available in documentation prepared for other purposes; some of it can be provided by a team

**Table 4.4** Illustrative HAZOP Matrix for a house water system.

| **Parameter (to right)<br>Guideword (below)** | **Flow** | **Temperature** | **Pressure** | **Level** |
|---|---|---|---|---|
| High | X | X | X | X |
| Higher | X | ? | X | X |
| Low | X | X | X | X |
| Lower | X | – | X | X |
| Reverse | X | – | – | – |
| As well as | X | – | – | – |
| Other than | ? | – | – | – |

X = Definite combination.
– = No realistic combination.
? = Possible combination – do not reject it without study.
Other guidewords:

- Access
- Maintenance

member when an issue arises; but some will have to be collected and put into a suitable format in readiness for the HAZOP study. The information should be 'firm' or 'frozen', i.e. not subject to continuing development and change. After it has been assembled and reviewed, a selection should be circulated to team members in advance of the first meeting. The team leader should consider the breakdown of the operation into stages for study by the team, select the guidewords to be used and prepare a draft list of parameters. A continuous process is normally divided into sections of hardware that are analysed in turn, working downstream through the process until all of the selected process has been considered. A batch process is divided into stages such that when all have been analysed the complete operation has been covered. Due to the focus on the stages of the batch it is likely that key items of the hardware, such as a reactor, may occur in several stages but in a different state or with a different activity each time. Each step must be given a clear and succinct word picture which is included in the worksheets.

It is also important that the worksheets follow a standard lay out. This will greatly assist the QA and auditing of the study. HAZOP: Guide to best practice [4] discusses this layout and the audit process.

## Analysis during the meetings

At the first meeting an overview should be given of the whole project to ensure that team members have a common understanding. A section or stage of the process can then be taken, starting with a detailed description of the equipment and operation from which a design intention can be developed. The search for deviations from this can then begin. It may be centred around guidewords, taken in turn, or around parameters.

The guidewords usually used are illustrated in Table 4.2.

These are combined with a parameter of the system to give a deviation. If the combination is physically meaningful then the team are asked to suggest causes. Where a possible cause is found then the consequences should be evaluated, using the team members' knowledge and experience to describe this. Unless the cause is extremely unlikely to occur or the consequences are trivial the team proceeds to consider the existing safeguards for this occurrence and decides if these are adequate. A record is made of the envisaged sequence and an action is generated if the safeguards are deemed insufficient or if any aspect cannot be described with confidence. The action is normally placed on an individual, often a team member, and the response will come back to the team for evaluation and confirmation. The process continues until all causes for all deviations (all combinations of guidewords and parameters) have been considered. The team can then move on to study the next stage in the process. Fig. 4.1 shows the pattern of working for the guideword-first approach.

Two points are worthy of emphasis. Firstly, in contrast to the list of guidewords, it is not possible to give a list of parameters. These must be thought out by the team members for the particular process being examined. However, Table 4.1 gives some examples.

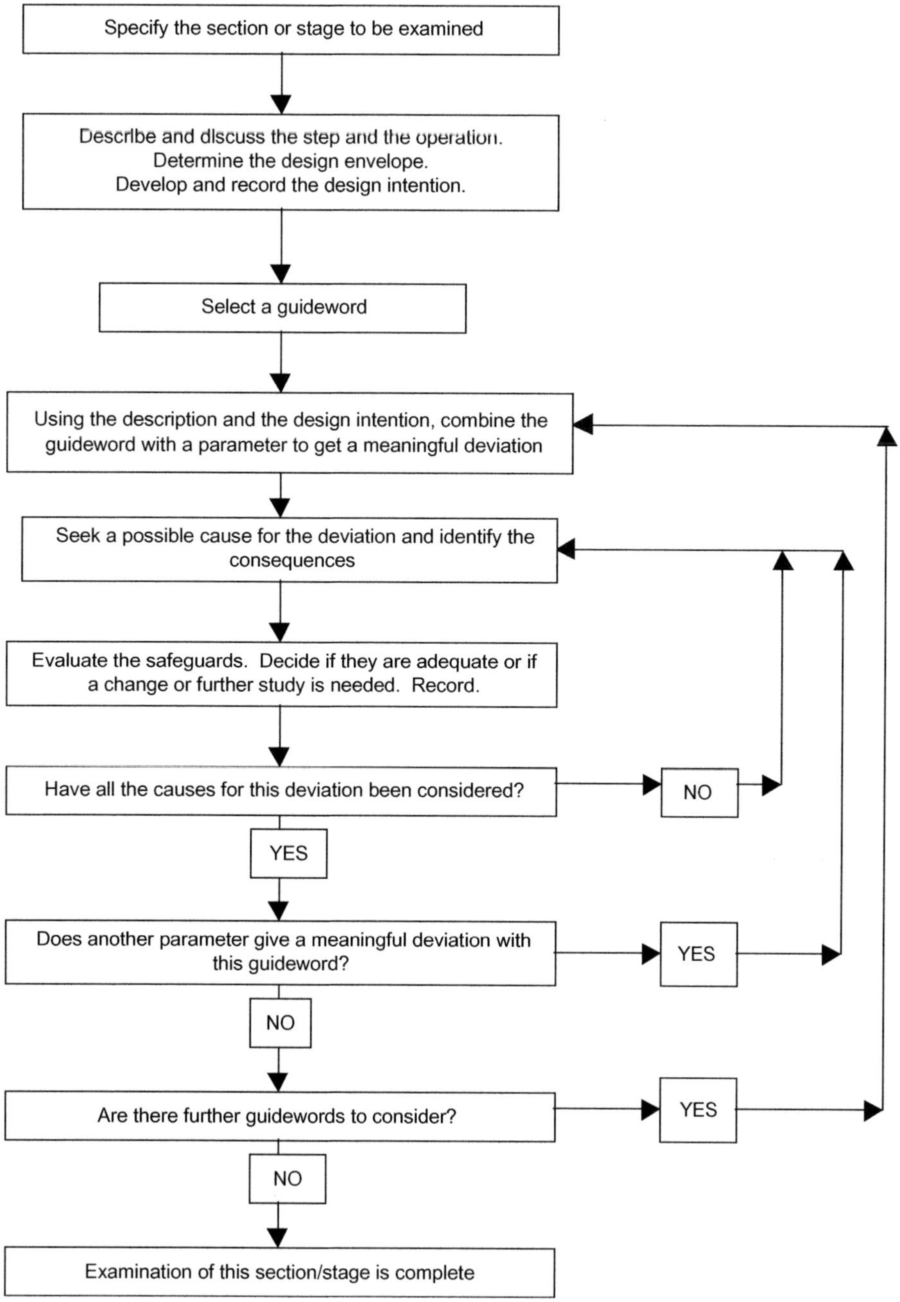


**FIG. 4.1** Flow diagram for the HAZOP analysis of a section or stage of an operation. The guideword-first approach.

Other case-specific parameters could include:

- Crystal size
- Crystal mass density
- Crystal properties (wetness and morphism)

In drilling other parameters may apply:

Fluid density

For Human Factors other parameters may be relevant:

- Confusion
- Misinterpretation of an instruction

Secondly, it should be understood that the main responsibility for suggesting parameters and deviations, for identifying possible causes, for analysing consequences and safeguards and for taking decisions rests with the team members. The primary tasks of the leader are to ensure the HAZOP method is being applied well, that the exploration of problems is to sufficient depth and that all necessary aspects are being recorded.

### The report

The records should be designed to meets the needs of the study as well as any subsequent uses. Later uses may include auditing, as a starting point for further hazard studies, as part of a safety case or during an incident investigation. Within a study there will draft reports and action notes issued after each meeting and both summary and full final reports. All of the working documents used in the study should be included or referenced in the full final report. A good report will cover the methodology used, attendance, section breakdown, process description and each design intention. It should show which guidewords and parameter combinations were used and the causes, consequences, safeguards and recommendations for each deviation considered. Action responses should be incorporated into the final report. It is commonplace for the records to be created directly on a computer using specialised HAZOP recording software. The actions can then be abstracted and tracked to completion.

A study should be formally signed off when it is completed and it should be clear who receives the report and has responsibility for implementing the recommendations made by the team.

## 4.3 Resource requirements

### Manpower

This method is very demanding of both personnel and time. On average there will be 6–10 experienced staff present at a HAZOP meeting and it is likely that, even with practised users, each section will take 1–2 h (and occasionally longer) to complete. Apart from small modifications, few studies can be done in a single session whilst for a major project the meetings will take up many weeks.

### Data

Detailed design data, process conditions and sufficient detail of the control and operating procedures are essential. If the team cannot create an accurate conceptual model of the eventual system they will either miss potential problems or misjudge their likelihood or consequences. Also the data must be firm with any changes made independent of the HAZOP study being done within a strict management-of-change procedure.

Due to these requirements any projects subjected to HAZOP study should be carefully selected to justify the effort that will be needed. Good planning and integration with other hazard studies and the overall project management will benefit the HAZOP study. The method should be applied with discretion.

## 4.4 Timing

A HAZOP study can only be done when detailed design information is available. To maximise the benefits it should be done when it is still possible to make both hardware and software changes to the plant and process. This determines the ideal time in the project as immediately upon completion of the detailed design work and when a minimum of firm commitments have been made for the acquisition of plant items. For modifications the study should precede their implementation whilst for an existing process a study may be done at any convenient time. It is becoming a requirement to carry out a new HAZOP after about 5 years when there may be a deviation from the original design intent used in the previous HAZOP.

## 4.5 Advantages, disadvantages and uncertainties

### Advantages

- The method can identify operating problems as well as hazards.
- Due to the structured approach there is a high probability of identifying the hazards.
- A wide range of hazards can be assessed, including chemical, mechanical, electrical, control and human interaction.
- New and novel processes can be investigated.
- The team gains a deep understanding of how the process is likely to operate.
- Much better operating procedures can be written after the study.
- There can be a financial payback from a faster start-up, fewer operating problems and increased reliability.

### Disadvantages

- The high resource requirements, both in manpower and data.
- The need for a multidisciplinary team and an experienced leader.
- The need to carry out the study during a narrow window in the project life.
- Great care must be taken if a plant or section is treated as a repeat of one studied previously as two systems are seldom truly identical.

### Uncertainties

- No team will find all possible problems. The proportion depends on the skills and experience of the team; the rigour and imagination with which the technique is used and the data provided for the study.
- The model assumed by the team must be representative of the final operating plant. Whether this is so depends on the control of later changes, the way the process is actually operated and how the plant is maintained. New hazards, which could not have been foreseen by the team, may easily be introduced if there is poor management.

## 4.6 Applications

Although the method was initially developed for use on new designs it quickly became used for modifications and for reviews of existing plant. Both continuous and batch processes can be studied. There are good reasons to carry out a 'new HAZOP' after a period of time to ensure that all of the interactions from modifications have been fully analysed. The same could be true where process knowledge has developed and should be incorporated into the knowledge base. Computer-controlled processes can be examined, as can contract or repeat designs, plant commissioning and even small-scale pilot plant or laboratory operations. In principle it can be applied to any operation for which a conceptual model can be created and a design intention formulated. The method of seeking deviations is then likely to uncover potential problems. As shown in HAZOP: Guide to best practice III [4], HAZOP can be applied to computer systems but the parameters must be adjusted. Likewise when analysing a Procedural HAZOP, which will include some Human Factors HAZOP: Guide to best practice III [4], a slightly different approach is required.

Many industries have benefited from using the method, including chemical, petrochemical, oil and gas both on- and off-shore, pharmaceutical, general manufacturing, food and water and transport.

Note that the preliminary HAZOP was mentioned earlier. This allows some of the more common findings of a HAZOP to be eliminated in the FEED stage before it enters the detailed design and HAZOP. This essentially forms a check list as show HAZOP III [4].

At the end of Safety Study 2 the level of development of design and P&IDs would be 'Approved for Design' (AFD), i.e. fairly basic but with the main process features in place.

The main features should have been added but the finer ones will not. It would be appropriate to examine the AFD diagrams for the more common errors using a form of check list. This could include:

1. Is the mass balance measured and achieved by the controls?
2. Are the main parameters of P, T, F and L adequately controlled and are there adequate diagnostics to assess problems?
3. Do any control parameters require independent verification?
4. Are piping materials of construction in compliance with codes? Does the spec reflect possible transients during pressurising and depressuring cycles?
5. Does this also apply to other associated equipment?
6. Are spec breaks in the correct place? Where fluids passing through let down systems are, is there a need to include a section of low-temperature piping downstream of the let down?
7. Are high- to low-pressure interfaces identified and given the correct over pressure treatment? Should the LP side be designed to the pressure of the HP side?
8. Has over pressure protection been applied as appropriate?
9. Are safety instrumented systems fitted where a parameter may result in a potential hazard?
10. Do safety instrumented system shut downs have prealarms?
11. Can safety instrumented systems be trip (function) tested without a process shut down?
12. Are the standards of isolation appropriate for the risks?
13. Are vents and drains required for maintenance shown on the P&ID?
14. Can all equipment requiring routine maintenance be cleaned and removed safely?
15. Is lagging appropriate to the piping codes?
16. Are non-return valves fitted where appropriate (pump discharges is a given)? See 14, 21 and 22.
17. Are there traps in vessels that may need special features for entry?
18. Are there any possible settling out points in piping or equipment which may need treatment?
19. Does the over-/underpressure relief indicated look 'adequate'?
20. Is the design of the relief collection consistent? Can incompatible materials mix in the system?
21. Do the vents go to a safe location?
22. Do the drains go to a safe (contained) capture system?
23. Is the design of the vent collection consistent? Can incompatible materials mix in the vent system?
24. Is the design of the drain collection consistent? Can incompatible materials mix in the drain system?
25. Are line slopes shown and are they appropriate?
26. Are there any pockets in the lines which may require drains?
27. Are liquid capture bunds required anywhere for S or E requirements?

**28.** For piping sizing use a rule of thumb of 30 m/s for gases and 2 m/s for fluids. Pump suctions should be treated specially.

It is to be assumed that issues such as erosion and corrosion will have been covered in HS1. In reality as with any HAZOP, these questions can be adjusted for a case-specific design.

## Other readings

A useful analysis of papers on HAZOP is given by J. Dunjóa et al., Hazard and operability (HAZOP) analysis. A literature review, J. Hazard. Mater. 173 (2010) 19–32.

Ref. [4] gives worked examples of Batch and Steady-State Process plus a Procedural HAZOP.

## References

[1] Guide to Hazard and Operability Studies, Chemical Industries Association, UK, 1977.

[2] Guidelines for Hazard Evaluation Procedures, AIChE Centre for Chemical Process Safety, USA, 2008.

[3] BS EN 61882, Hazard and Operability Studies (HAZOP Studies), 2001 (Application Guide IEC 61882 Guide for Hazard and Operability (HAZOP) Studies).

[4] F. Crawley, B. Tyler, HAZOP Guide to Best Practice, third ed., Elsevier, ISBN: 978-0-323-39460-4, 2015.

# 5

# HAZID

Chapter outline



## 5.1 Precautionary comment

HAZID evolved from HAZOP in the early 1990 and was given its name by the author. The origins are to be found in the references. Since then Hazard Identification has been contracted to HAZID which leads to confusion.

## 5.2 Definition

A systematic review of the possible causes and consequences of hazardous events [1]. This is a simple but accurate definition.

It has been a particular form of Hazard Identification widely used offshore and has proven most useful at Hazard Study 2 (**F**ront-**E**nd **E**ngineering **D**esign) where 'risks' can be designed out relatively cheaply and also construction.

A Guide to Hazard Identification Methods. https://doi.org/10.1016/B978-0-12-819543-7.00005-7 

## 5.3 Description

HAZID originated from HAZOP and was originally known as '3D HAZOP' [2] as the main objective, at that time, was to determine the interaction of potential 'risks' in a three-dimensional space typical of offshore installations. After 4 years of evolution it became known as HAZID. While HAZOP is '***cause***' driven HAZID is '***consequence***' driven. As with new techniques it took time and usage to develop it to the full potential. Initially it was based upon a HAZID matrix, which had three headings:

Effect,
To/On,
Cause

By permuting each entry from column 1 with one from both columns 2 and 3 various risk scenarios could be proposed. It was then necessary to assess, through quantitative risk assessment, if the scenario was high or low consequence/frequency. (This had parallels with HAZOP.)

In Ref. [2], Table 5.1 with three columns was used to define the issues.

These columns, which were derived for the offshore O&G process, are for illustrative purposes and are not complete.

In Ref. [1], the original table was rearranged as Table 5.2:

In effect the To/On is column 1 in Table 5.1 is the same as the cause in column 2 of Table 5.2.

The structure of the matrix is a case-specific study used to flush out issues by permutations of an entry in each box. It can be seen that Tables 5.1 and 5.2 are virtually the same.

With experience in the use the table was shortened to the following:

**Table 5.1** Original matrix of matrix in Ref. [2].

| Effect | To/On | Cause |
|---|---|---|
| Mechanical damage | Structures | Crane movement |
| Asphyxiation | Control room | Dropped loads |
| Loss of control | Equipment | Explosion and missiles |
| Distortion | Instruments/controls | Fire (lack of protection) |
| etc. | Personnel | etc. |
| | etc. | |

**Table 5.2** New matrix as derived from usage.

| Generic hazard | Cause | Effect |
|---|---|---|
| Loss of containment | Design/construction | Fire/explosion |
| etc. | etc. | etc. |

### To/On

*Effect*

1. Structures
   Fatigue
   Corrosion
   Earthquake
   Impact
   etc.
2. Process/pipelines
   etc.
3. Services
   etc.
   Subsets
   etc.
4. Safety systems
   etc.
5. Communication
   etc.
6. Safe haven
   etc.
7. Coordination
   etc.
8. Others

This is easier for the Facilitator to construct 'scenarios' with the client from background experience but must be used with care.

## 5.4 Resources

### Preliminaries

The preliminaries are the foundation of all Hazard Identification Studies. The Facilitator plus another as appropriate must liaise with the 'project manager' to establish the ground rules:

1. Scope

   The scope should include that which is in and that which is out of the remit of the study. If this is not carried out the study may stray into areas, which are not requiring study with the inherent waste of time.
2. Terms of reference

   This should include the detail to be used and whether the actions should be ranked. It should include the delivery date and format.

**3.** Matrix (see later)

This is key to the study and should be a blend of the project knowledge as well as that of the Facilitator. The construction of the matrix may involve more than just the Facilitator and Project Manager. In complex issues the Facilitator can probe areas, which may not have been considered previously and also gain a deeper understanding of the issues. Time spent on the construction of the matrix will be well rewarded in the study.

## 5.5 Manpower

Typically the core team would contain a Facilitator, a Scribe (as HAZOP), a Process Engineer, a Project Engineer, Mechanical Engineer and a Safety Engineer.

Depending upon the project phase and project nature, which may be mechanical or civil in origin, which should include, as examples:

Construction Engineer, Civil Engineer, Mechanical Engineer, Fire Protection Engineer, Chemist, Structural Engineer, and if necessary Transport. The skill base has to be chosen carefully and could change as the study proceeds through the flow sheet. The addition of a new team member (or one leaving) has to be timed correctly such that that person is fully *au fait* with the process when so required.

### Data

Typically (as with HAZOP) there are key drawings:

Plot Plans with elevations, Process Flow Diagram giving the nature of the materials handled in each section of the process pressure, temperature and composition for a Process Study. Materials Hazards Data Sheets, Materials of Construction, Protective Systems—passive and active, Procedures. In other cases the drawings could include detailed structural and fabrication drawings and tests carried out on construction materials. It may be necessary to include stress and fracture analysis studies. Each discipline should be able to draw up their own needs.

In addition a 'matrix' of ***generic hazards***, **causes** and **effects** must be drawn up with the project and probably an experienced Safety Engineer.

Do not underestimate the time as it is critical to the study success.

## 5.6 Timing

Ideally the HAZID Study would start at Hazard Study 1 and then be refined at Hazard Study 2. Equally it would be used to study the final shut down of the process and its demolition.

HAZID can be carried out at all project phases as defined in Process Hazard Studies (Chapter 3).

Ideally it fits best in Review 2 as it is the precursor to Quantified Risk Assessment leading to the Safety Case. It is also key in Study 7 (Demolition) and will be part of the Demolition Safety Case.

## 5.7 Advantages, disadvantages and uncertainties

### Advantages

- It is possibly the best tool for analysing the process at Hazard Study 2 leading up to the first full Quantified Risk Assessment.
- It fits in very well to the abandonment and demolition of process plant.
- The study will produce novel options for change.
- It is flexible and can be used at different phases of a project definition.
- It utilises the experiences of operational staff.
- It is relatively quick and avoids repetitive analysis.
- It can identify low-frequency high-impact events and so fits in well with QRA.

### Disadvantages

- The study is mentally exhausting.
- The team structure may have to change during the study process. The timing of the arrival of new member(s) and the departure of ones who cannot contribute to a new phase requires to be handled with care. There is a 'learning curve' of about 1 hour.
- The matrix requires development for each installation and may omit some hazards.
- Its benefits depend on the experience of the leader and the knowledge of the team, as with HAZOP.

The last item is true for any study but if the preliminary meeting functions as it should do there should not be a problem.

### Weaknesses

- The generation of the guidewords may omit a significant event.
- The success is dependent on the skills of the team.

### Uncertainties

- The matrices require a detailed understanding of the failure modes of people and equipment and are case specific not generic.
- All actions require a form of QRA by which they will be prioritised.
- Actions defined at the study meeting may require to be changed as a result of the QRA—this does not diminish the technique.

## Worksheets

The following worksheet is shown to indicate the detail required and is not to be treated as a real case study.

The header for the worksheets will include the date of the study, attendees and affiliation, drawings used, revision number and date plus any other data sheets and dates used during the study. Basically this is who was there and what information was accessed. This is as in any HAZOP (Table 5.3).

The risk assessment must be carried out AFTER the meeting when all of the data will be available. The column describing the scenario will assist the risk assessor. The risk assessment can be carried out on a separate worksheet with the date and assessor and copied into the final worksheet. The risk assessment worksheet (QRA) is important for the Quality Assurance and may become part of the Safety Case Package.

**Table 5.3** Typical worksheet for HAZID.

| Ref. no. | Area | Effect | Cause | Consequence | Mitigations and controls | Scenario | Risk[a] Profile F&M | Ref. no. Actions[b] Actions on[c] |
|---|---|---|---|---|---|---|---|---|
| 1.1 | Control centre | Fire damage | Torch fire from adjacent piping | Persons trapped | Passive fire protection | Torch fire from ruptured pipeline. Leak caused by impact from dropped/ swinging loads | To be filled in outside the meeting<br>$F = 10^{-3}$<br>M = Risk 3 | 1.1.1<br>Forbid lifts over the equipment<br>J Soap<br>1.1.2<br>Review vulnerable areas of control centre<br>J Bloggs |
| 2.1 | Storage | Acid corrosion | Leaking joints on vessel above storage | Release of toxic materials leading to injury of personnel | Routine inspection | Vessel F1 develops a leak of acid which falls on vessel F2. This results in the release of toxic fluids | | 2.1.1<br>If FEED reverse the location of the two vessels or redesign the lay out<br>2.1.2<br>I non FEED consider the need for a corrosion resistant drip tray under F2<br>A Scouser |

[a] To be filled in outside the meeting.
[b] To be carried out outside the meeting.
[c] To be filled in outside the meeting.

Two other simple examples are:

1. Effect—fire damage—on/to—power cables—by pump fire or pump seal failure

Actions:

   a. Consider the need for secondary seals
   b. Consider the need to relocate the pump away from the cables
   c. Consider the need for passive fire protection for the cables

2. Effect—poor maintenance access—on/to heat exchanger E1111—exchanger cannot be removed or may damage other equipment when being removed

Actions:

   a. Carry out a mechanical handling review and then review the need for a different layout.

## 5.8 Methodology

The first step is to select a team with the appropriate skills to that particular study. As an indication some teams in the offshore industry had sub-sea engineers, divers, drillers, construction engineers and logistics. The team selection is as important as the *matrix*.

As the HAZID study is case specific so also may be the team membership.

The room selection is very much as with HAZOP (see Chapter 4)

Scope
Definition
Describe the matrix

The starting point (as with many Hazard Identification Techniques) starts with the *matrix* as defined by the dialogue between the Facilitator and the Project.

The matrix will be case specific. Unlike HAZOP where the parameters and the guidewords are fixed by the process in HAZID the matrix is fixed by the objective of the study. This is illustrated in Table 5.1 for a generic case and the worked example for a specific case.

No two matrices will be the same. This emphasises the need develop the matrix for any specific discipline and to spend some time with the project developing the matrix.

The Facilitator must now select an area, as shown above by the matrix, and ask what might cause damage to it or impair its function. One example that usually arises with the addition of temporary equipment is the loss of function of fire detection and poor illumination. Equally it could impair the means of escape and maintenance access. Another might be the impact of loss of containment on adjacent equipment. As is shown in the disadvantages the skills of the Team and Facilitator are fundamental to the success of the study. The Facilitator has to encourage discussion of the lead topic in the matrix and to challenge replies where they are poorly explained or defined. The worked example shows this in some detail

The dialogue process, Facilitation, is in line with that of an Audit. It is mentally more tiring than HAZOP and requires skills in '*eliciting information*'. The task is to pose a pointed question relating to one of the items in the matrix and then to use that answer to open up the discussion. It is very much a mental challenge as the Facilitator is trying to interpret ideas which are often incomplete. Do not underestimate the mental fatigue in this process.

One feature of all studies is that initially the team, who may be part of the project, will be sceptical of the process and possibly reluctant to have their design challenged. This is quite normal but within an hour the team will realise the benefits of the study.

Finally as the outcome of the study will include some high-risk but low-probability events ($10^{-4}$ per year are possible with a trained team and Facilitator) a detailed QRA to assess the priority of the actions arising. This is best done outside the meeting to avoid loss of focus in the study and is typically about 15 min of work for each finding (unlike a HAZOP).

## Appendix

The following is a worked example of HAZID in action. It is based on a real event witnessed by the author which would have been averted if HAZID had been used.

### Task to be carried out

A rectangular structure about 10 m high and 1.5 m at a square base was to be fitted on the side of a Marina near to a large harbour. This structure was to carry Christmas lights and various Tableaus. Access to the Marina was limited to small, light bridges over the inlets (leads) which could be raised for access of yachts. Mobile cranes could not be used due to bridge width and load limits.

It was decided to use an Aerospatiale AS350B2 helicopter on the freely accessible side of the Marina side to lift the structure which would be taken over the bridge in small pieces and assembled on the side of the marina.

The tower was assembled and a lifting line was attached to it. The tower pivoted on two pins at the base and once raised two more pins at the base would be inserted and guys attached. The lifting line was to be disconnected automatically from the helicopter by the pilot once the tower was properly located.

It is arguable that an 'inherently safer' approach would have been to locate the tower elsewhere. This would have used a large part of a car park, created security issues and was rejected.

The HAZID must start with the matrix (Table 5.4).

An alternate, preferred, matrix derived from usage is given below. Each structure or matrix is equally acceptable.

Helicopter crash

Loss of control
Wind gusts affecting control close in to the structure
Asymmetric load line snagged on a skid
Engine failure
Loss of power or fuel
Is the pilot adequately trained and aware of the 'emergency procedures'?

**Table 5.4** Helicopter crash HAZID matrix.

| Generic hazard | Cause | Effect |
|---|---|---|
| Disengage does not function | Poor fitting/testing<br>Unreliable | Helicopter left connected to tower—potential crash |
| Pilot error | Lack of experience of this type of lift | Potential crash |
| Lifting hook over loads Helicopter | Helicopter cannot lift the load or is damaged in so doing<br>Can the helicopter be controlled adequately? | 1. Snatch loads<br>2. Unusual off set loads<br>3. Damage to Helicopter<br>4. Potential for crash |
| Recovery from a failed lift. How is it to be carried out? | Lifting line does not disengage correctly | Possible Helicopter crash |
| Helicopter crash | Loss of control<br>Wind gusts while lifting<br>Asymmetric load<br>Engine failure<br>Loss of power/fuel | Needs for emergency services |

Fire following a crash

Emergency services availability—fire & ambulance

Helicopter overload

Can the helicopter be controlled adequately?
What is the maximum safe off centre loading?
What calculations are available?

Lifting hooks and cable overload
Function testing of disengagement link

What happens if the cable cannot be disengaged?

The full HAZID worksheet will not be produced only an indicative selection of ideas as well as some dialogue.

**Dialogue** (**F** = Facilitator **T** = Team Member)

**F** What is the key safety feature of this lift?
**T** The lifting lug and the release mechanism.
**F** How can you **guarantee** that the lifting line WILL disengage when required?
**T** This is a standard procedure! —— If it does fail the helicopter will be attached to the tower!
**F** Are there any particular features of this lift which might create a failure of the disengagement system?
**T** We cannot think of any. (This might warrant further discussion)
**F** Should the emergency release be proof tested before take-off?

**T** I suppose we should test it before take-off!
**F** Can you be sure that the proof test will replicate the real disengagement?
**T** We will check this in more detail
**F** Record as a review item
**F** What is the recovery procedure? Can the helicopter land to one side?
**T** The lift line is only intended to be 5 m long!
**F** What is the limit? If it was 15 m long could the helicopter land safely?
**T** Interesting point!
**F** Record as a review question.
**F** Is the pilot trained and experienced with this type of lift? Has he a proven track record? Can you provide evidence if there is an upset?
**T** Yes! He is the best!
**F** Record this and put the experience record into the project dossier.
**F** Have the loads on the helicopter fastenings been examined by a stress analyst? Are there any limitations to the snatch or offset loads? Is there a certificate of air worthiness for this operation?
**T** This is a standard routine!
**F** I repeat the question. Are there limitations to the loading? Can the helicopter be controlled in an offset load? This should be an action!
**T** I suppose so!
**F** Record
**F** What emergency services should be present?
**T** Fire and ambulance!
**F** Good but will there be foam?
**T** I suppose so!
**F** Record

This dialogue has been written in the form chosen as it is not unusual for teams to be very defensive initially—'What do you know about it in any case?' Once the Team gains the confidence of the Facilitator and vice versa the style will become more relaxed.

There was NO emergency plan. The lifting line did not disconnect. The helicopter descended to the point that the lifting line was clear of the rotor and a rigger jumped up to release the line manually. This created an asymmetric load and the helicopter hit the structure and crashed.

*Note*: One of the key words produced before the accident investigation was produced was *asymmetric load*. This appears to have been prophetic.

No one was seriously injured. The author witnessed the event but did not see the rigger jump to release the line as the whole operation was an accident waiting to happen and this author turned his back to the fiasco.

CAA Occurrence Number 11/5227, WR 12/SAJ/245.

## Helicopter lift HAZID

Date 1/12/2019

Attendees: AB (facilitator), BC (Scribe), A Scouser (Project Coordinator), Abe Baker (Helicopter Company)

**Worksheet 5.1** HAZID on tower lift.

| Ref. no. | Area | Effect | Cause | Consequence | Mitigations and controls | Scenario | Risk[a] Profile F&M | Ref. no. Actions[b] Actions on[c] |
|---|---|---|---|---|---|---|---|---|
| 1.1.1 | Lift lug | Damage to fuselage | No certification | Loss of control | None | Lift Lug results in damage to fuselage and loss of control | See Action | 1.1.1<br>Verify that there is a certificate of air worthiness for this operation<br>A Scouser |
| 1.1.2 | Lift lug | Damage to helicopter<br>Poor design | Overload<br>Release not designed proper | Damage to helicopter No release achieved | None | Snatch load damages Helicopter or disengage does not work | | 1.1.2.1<br>Record any calculations on the load on the helicopter<br>A Scouser |
| | | | No function testing | Helicopter remains tethered | None | | | 1.1.2.2<br>Record the assessment of the disengage performance<br>Abe Baker |
| 1.1.3 | Control | Lift Line does not separate | | Helicopter crash | None other than the selection of a day with still weather | Helicopter pilot attempts the separation but the mechanism does not operate as intended | $F = 10^{-2}$<br>$M = 4$ | 1.1.3.1<br>Ensure that there is a fully developed function test for the separation mechanism which replicates the real event.<br>Implement the test prior to the lift and record it in the data sheets.<br>Abe Baker |
| | | | | Impact on structure or other causes of a crash | None at present | | | 1.1.3.2<br>How do you handle this scenario? Can an emergency procedure be devised?<br>Abe Baker |
| 1.4 | Wind | Helicopter difficult to control—potential crash onto tower | Gusty wind | | | Helicopter driven onto structure by a gust and crashes | $F = 2$ (uncertain)<br>$M = 4$ | 1.4<br>Is a 15 m tether acceptable?<br>A Baker<br>Examine the best weather window for the lift to avoid gusts.<br>A Scouser |
| 1.5 | Site emergency | Crash and subsequent fire | Loss of control | | | Helicopter crashes due to loss of control: offset load, wind gust | $F = 2$ (uncertain)<br>$M = 4$ | 1.5<br>Arrange for a standby fire tender with foam<br>A Scouser |

[a] To be filled in outside the meeting.
[b] To be carried out outside the meeting.
[c] To be filled in outside the meeting.

Data: Provisional operating procedure (date 1/2/10). Site location and access limits. Wind distributions and speeds (from local airport).
Actions would be:

1. Determine the safe operating wind envelope.
2. Produce an emergency abort plan.
3. Function tests all emergency systems under realistic working conditions.
4. Have a fire tender on standby with foam facilities.
5. Record the experience of the pilot.
6. Record the calculations on loading plus the design of the release mechanism.

# References

[1] J. Spouge, A Guide to Quantitative Risk Assessment in Offshore Installations, DNV–GL, Horvik, Norway, ISBN: 1-870553-365, 1999.

[2] F.K. Crawley, The application of hazard and operability study techniques to the design of offshore installation topside facilities, in: Offshore Safety Conference, IBC Technical Services, London, October 1987.

# Further reading

[3] F.K. Crawley, et al., Can we predict potential major hazards? Major hazards onshore and offshore, in: IChemE Series 130, IChemE Rugby, Manchester, 1995.

# 6 Task analysis

Chapter outline



## 6.1 Definition

Task analysis is a structured analysis of a task. It is done by breaking down the process into detailed steps so that each part of the sequence can be examined to identify what might go wrong and what can be done to prevent or mitigate the consequence.

## 6.2 Description

Task analysis analyses every step of the selected operation. This could be a process-related task, a maintenance task or an inspection task. The task is first analysed to show the precise sequence of steps that are needed, including what the operator does, what hardware changes are made or will occur in order to achieve the final goal. The analyst then considers each step and asks what could go wrong with this step or if it could be omitted or done in the wrong order and then how any problems could be eliminated/mitigated. In the cases of maintenance and inspection there will be an interface with the operations group who will prepare and deliver the equipment. It is usually applied to safety critical steps in a process and is especially useful in analysis of operating procedures.

It is of note that the use of this technique in a home would significantly reduce the number of unnecessary accidents!


A Guide to Hazard Identification Methods. https://doi.org/10.1016/B978-0-12-819543-7.00006-9

## 6.3 Resources required

*Manpower*: For a simple task analysis the resources could be as low as one or two experienced persons who have an understanding of and training in the method and of that particular task. Ideally they should be of the same discipline but occasionally it might be advantageous for a process representative to be part of the team when there is a particular interface such as cleanliness for maintenance or inspection and on-line testing. In the case of a routine task the analysis could be carried out by the supervisor and then recorded in the operating procedure or permit to work. However a major study such as an underpressure break-in will require engineering experience and a team approach leading to some form of Quantified Risk Assessment.

*Documentation*: As a minimum there should be a written definition of what should be achieved and how it might be achieved. A check list may also be appropriate.

## 6.4 Timing

The timing will be dependent on the task and preparation but as a minimum it should allow time for changes in the procedure and any training and supervision that might be required. It is possible that new equipment may have to be purchased so this has to be born in mind.

## 6.5 Advantages, disadvantages and uncertainties

### Advantages

- It uses the skills of persons who have experience with that task.
- It is tailored to the individual task and the requirement of that task.
- It can use a simplified Hazard Study format.

### Disadvantages

- It may rely on experience that is not relevant for the particular task or the experience may not be readily available.
- There is a temptation to rely totally on the experience and so lose some structure in the analysis when it might be more appropriate to examine simpler (inherently safer) methods.

### Uncertainties

- There may be a tendency to assume that the task is routine when it is more complex than anticipated and might be out with the supervisor's competence.
- See the second point under disadvantages.

## 6.6 Application

There are many areas where task analysis may help identify problems and hazards. Because it is time consuming it is usually restricted to operations recognised as having significant potential risks. A frequent use is in the preparation and issue of a permit to work.

Suitable procedures for study include:

- A change to an operating procedure
- A maintenance procedure
- An inspection procedure
- A test procedure such as function testing an instrument

Above all task analysis is appropriate to simple and routine tasks, not major projects. In the case of new or major changes to a procedure, including significant interface issues, it is more appropriate to adopt a Hazard Study approach.

### Operation of task analysis

Task analysis asks the questions posed earlier to a selected step which leads to a Risk Assessment, namely:

- What are the issues with that step?
- What is the order of each step? (Hierarchy)
- What can go wrong during that step?
- What can prevent or mitigate the effect?
- Should the procedure be modified or abandoned?

It is now a case of taking each step as specified in the procedure and asking these five questions in turn. The time required may only be 10 or 15 min, but the warning about the interface must be taken seriously.

Some historic examples of failure at the interface between process and analysts are:

- Trapped fluids in equipment delivered to a maintenance workshop.
- Poor planning for the removal of a power recovery turbine shrunk fit on a compressor shaft. The heating process was not clearly thought through.
- Poor understanding of the air flow regime in a vessel prepared for inspection.
- History notes that fires are a common feature of structural modifications in buildings. When working with gas cutters, welding, or tar boilers has attention been paid to local flammables, fire prevention, fire protection and fire attack in an emergency? This is exampled by the fire at Notre Dame—Paris April 15th 2019 and more locally to the author during roof work on a hotel in Ayr—Scotland—June 14th 2019.

## Hierarchical task analysis

While it is implicit in the analysis it should be emphasised that the order in which the sub-tasks are carried out and the missing sub-task may be critical. This is called Hierarchical Task Analysis. In some respects this is a 'critical path planning' (CPP) process. As an example it would be necessary to displace toxic materials by nitrogen or water before introducing air so as to avoid the creation of a potential explosive system. Of equal importance is that if a step is not completed correctly there should be a retreat position where further actions can be considered. (In a Procedural HAZOP this would be the use of the guide-words 'proof of success'). These actions can be tabulated as a plan where each step is signed off when completed. Such a plan would be used when purging out a system. The complexity of purging/sweetening should not be treated lightly. It is easy to recontaminate a sweetened section if the order is not ideal. This requires a very carefully defined plan with the order of purging with the position of the appropriate isolation valves. Each step has to be done in order, the results recorded before the next step can be started. Each step must have a clear reference for success and a pause condition set if not successful. The success step, with any analytical results, should be recorded and signed off.

Some possible guide words are as follows:

Parties involved
Timing
Sequence (order of operations)
Communication inter and extra the party (visual, verbal or written)
Plant settings at start and end of operation
Rate of change (of variable)
Verification of successful completion of each step
Hold and stop conditions

An operating procedure will have some elements of the above. There is also an element of Critical Path Analysis.

- The sequence (order or hierarchy) has to be defined with the criteria for success before the next step is attempted.
- The first step may be 'Isolate the item'.
- What are the criteria for success and how can this be proven? An example of this might be 'a leak-off test' or 'a pressure build up test' in the equipment.
- What action must be taken if there is a leakage through valves?
- If successful record the result and move to the next step in the sequence.

For routine tasks such as lighting a furnace the task can be pre-recorded into a worksheet and each step tracked to success with a signature.

Task analysis may prove more important for tasks such as:

- Entry into confined spaces.
- Inspections.

- Repair work on site and in a workshop.
- Function testing safety systems.

Within the example that follows there is hierarchy of actions that require to be carried out in the right order.

Due to the wide use of and range of applications for task analysis there is extensive literature coverage [1–4].

## Illustrative example

### *On-line cooling tower inspection—Task analysis for Permit to Work (PtW)*

It is necessary to carry out a routine, scheduled, internal inspections on a cooling tower internals to assess the cleanliness of the packing and to see if there is any damage to the packing itself or to the distribution troughs since this would lead to mal-distribution of the return water. The simple diagram below shows the unit and the process with sources of contamination (Fig. 6.1).

The cooling tower has an induced flow fan at the top rotating at about 120 rpm and a set of distribution channels over the packing. The distribution is by troughs in order to get a uniform distribution. The packing is made of wood. There is an inspection door below the fan but above the packing in the vertical section at the access way into the void between the fan and distributors. This is shown in the sketch on the LHS. With the fan shut down the cooling capacity is reduced to about 50% of the normal working flow. However in winter, when the air is colder, the impact on cooling can be relatively little and fans can be shut down as an energy saver.

Relevant material properties are:

- Hydrogen sulphide, $H_2S$: The workplace exposure limits (WELs), for $H_2S$ are 5 ppm [8-h time-weighted average (TWA)] and 10 ppm (15-min TWA)
- Chlorine, $Cl_2$: The workplace exposure limit for $Cl_2$ is 0.5 ppm (15-min TWA)

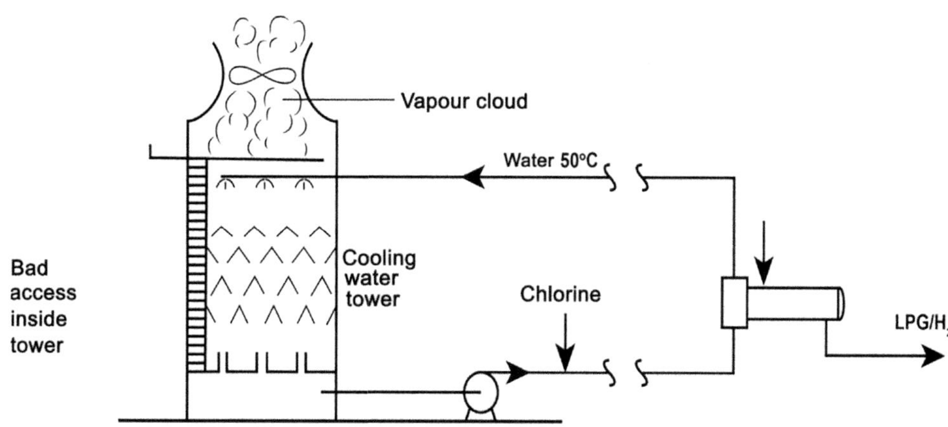


**FIG. 6.1** Simplified sketch of cooling tower. *Courtesy IChemE.*

- LPG, treated as propane, has a lower explosion limit (LEL) of 2.2% in air. An acceptable level would be no more than 25% of this figure.

The analysis can now be treated as a 'task analysis' leading to the issue of a PtW or as a mini-project to determine how it can be done safely. The issues that are recognised include:

- The fan creates a negative pressure of about 75 mm water over the door area of $2\,m^2$. This means that any entry through the inspection door is impossible with the fan running.
- There will be spray and mist above the distribution troughs when the fan is running. The mist prevents visual inspection and so the fan must be shut down.
- When the fan is shut down there will be about a 50% reduction in cooling capacity from that section. This means that the inspection can only be carried out in the winter.
- The surfaces will be very slippery.
- Cooling towers are notorious for the presence of the bacterium *Legionella pneumophila*. Cooling towers are chlorinated once per day to kill any pathogens in the water but is toxic to humans. Any chlorine has to be stripped out of the cooling water over a few hours of operation.
- There is always the risk of process leakage into the water—in this case it is hydrogen sulphide as well as flammable hydrocarbons. The hydrogen sulphide is the critical contaminant.
- The water is usually treated with chemicals to prevent (arrest) any corrosion.
- There will be some fine mist above the packing even when the fan is shut down.

The analyst must ask how these issues can be safely accommodation in a procedure.

- The fan must be shut down to allow entry. This means that the inspection can only be carried out at either low cooling demand or in the winter when the air is cold.
- The fan starter must be locked off to prevent start while the inspection is carried out.
- PPE must include boots which will cut through slime.
- Chlorination must be stopped 24 h before inspection so as to strip out residual chlorine. The valves on the chlorine drums need to be locked closed.
- The return mains must be checked for contaminants. This can be done by sampling or by venting the high points in the return mains. If contaminants are found then additional sampling of gases in the tower is required. (By design there will be a 'high point vent' so as to remove any air in the exchanger prior to start up.) The configuration is usually cold water in the bottom of the exchanger head and warm water out at the top. The vent will be in the vertical section of the top line leaving the exchanger.
- The chemical doping is a chromate, which is potentially toxic to humans. This must be discontinued but some residual will still be present in the cooling water.
- The mist may include chromates and Legionella. Chromates are of concern. Legionella does not affect healthy persons but this risk is unacceptable since it affects those whose immunology is challenged.

The analyst must also know the details of the Personal Protection Equipment (PPE):

- Boots must have good nonslip properties.
- The inspector should wear a fully fitted impervious suit with fitted gloves.
- The inspection should be carried out using breathing apparatus (BA)—preferably of the supply type to prevent ingestion of mists with chromates and Legionella.
- The inspector should wear a safety harness and line against slipping and/or distributor and packing collapse.
- A stand-by person should be on duty outside the inspection door to affect rescue in the event of an incident.

Pre-inspection

- The weather window or process cooling water load window should be identified.
- Tests for process leakage into the water should be carried out.
- Chlorination should be stopped 24 hours before inspection.

Preliminary tasks and checks

1. Notifying the users of the cooling tower that it is necessary to carry out an inspection and that there will be a loss of cooling capacity and that some adjustments may have to be made on coolers. This should be notified about a week in advance.
2. A dummy run should be done so that each person in the team knows exactly what is required of them and to demonstrate possible emergencies have been catered for.
3. Verification that the inspection team are trained in the use of breathing air.
4. Verification that there are reserves of air bottles.
5. Checking that leakage tests to the water have been carried out. (This can be done by venting the return main as it leaves the heat exchanger.)
6. Check that the chlorination has been stopped and that the chlorination valves have been locked closed 24 h before the inspection.
7. Issue the PPE.
8. Notify the users that the inspection will take place in the next hour and to be prepared to take actions.
9. Pause for the acceptance from the users of the cooling water.
10. Stop the fan and locking off the starter.
11. Issue the inspection permit specifying the Personal Protective Equipment (PPE).
12. Issue $H_2S$ dose cards for personnel entering the tower.
13. Issue the entry permit verifying that the chlorine and fan motor are lock off and that tests have been carried out for leakage.

Note that there is a hierarchy in this list.

1, 2 and 3 can and should be carried out a few days before the inspection
4 should also be carried out a few days before the inspection

5 must be carried out within 2 days of the inspection so that the users are ready for the event
6 must be carried out 24 h prior to the inspection
7–13 must be carried out on the day of the inspection.

*Clean up*

The inspection suits and breathing air masks should be hosed down to remove and traces of water-bearing chromates and possible Legionella. The inspection team should take a hot shower to wash off any residual contamination from the body as hot water will also kill any Legionella. The inspection team should then be on the alert for any flu-like symptoms for a week after the inspection.

*Reality check on risk assessment*

The risk assessment should consider how credible it is for a flammable or toxic atmosphere to exist within the cooling tower.

Firstly, to examine the likelihood of a flammable atmosphere. The tower is a square section $7 \times 7$ m and with air at about 40°C and a velocity of 1 m/s without the fan there will be a throughput of about 49 $m^3$/s. To achieve the LEL of the gas of 2.2% (v/v) requires a leak rate of over 1 $m^3$/s or around 6 t/h. Given the throughput of the plant this would have been found through routine mass balance checks. So not only is it extremely unlikely that a flammable atmosphere could occur there is the further safety factor that in a dense misty environment it is virtually impossible to ignite a gas as the flame front is usually quenched by the mist.

Secondly it is necessary to examine the toxic risks. At a high value of 1000 ppm $H_2S$ in the LPG the leak of 6 t/h would give a $H_2S$ level of around 22 ppm. Since the acceptable flammable level is only 25% of the LEL this gives $H_2S$ at no more than 5.5 ppm. Thus this too is an unrealistic eventuality. However, it is worth using breathing sets to prevent contamination by chromates and Legionella.

## References

[1] HSE Guidance on Operating Procedures, see www.hse.gov.uk/research/otopd/1999/oto99092.pdf.

[2] B. Kirwan, L.K. Ainsworth (Eds.), A Guide to Task Analysis, CRC Press, 1992.

[3] CCPS Guidelines for Preventing Human Error in Process Safety, Wiley, 2004.

[4] S. Mannan (Ed.), Lees Loss Prevention in the-Process Industries, fourth ed., Butterworth-Heinemann, 2012.

## Further reading

[5] The Loss Prevention Bulletins now on the Safety and Loss Prevention IChemE web site are full of examples of this failure.

# 7 Layer of Protection Analysis (LOPA)

Chapter outline



## 7.1 Definition

Layer of Protection Analysis (LOPA) is an analytical procedure that looks at the safeguards on a process plant to see if the protection provided is adequate for every identified hazard. It shows whether additional controls or shutdown systems are required by comparing the risk, with and without these additional elements, against a set of predetermined risk criteria. LOPA started life as an order-of-magnitude technique but has developed by using factors such as failure frequencies and probabilities of failure on demand, which have been recorded or estimated using well proven methods. This is particularly relevant for human factors, which are frequently initiating causes of hazards and are involved in assessment of alarm systems as a layer of protection.

## 7.2 Description

LOPA looks at the independent protection systems in a chemical plant and examines their actions when an unwanted event develops to determine their adequacy [1–4]. Typically the layers of protection, intended to provide defence in depth against a developing event, are:

- Good process (plant) design and engineering, mechanical integrity.
- The process control system (BPCS) and operating procedures.

A Guide to Hazard Identification Methods. https://doi.org/10.1016/B978-0-12-819543-7.00007-0

- Critical alarms and manual intervention. *Note*: this layer may be BPCS related and can include BPCS driven IPLs so a caution needs to be placed about the independence of this layer.
- Automatic SIS or automated ESD. *Note*: If an ESD requires manual initiation following receipt of an alarm it should not be included in this layer.

The next layers may act to reduce risk if they result in a safe outcome but in general they act to mitigate the scale of an event.

- Physical protection such as relief valves.
- Site emergency support and services.

Fig. 7.1 shows the 'onion' view of the LOPA layers.

The eight concentric circles are:

1. Process design.
2. Basic process control systems (BPCS) and operating procedures.
3. Alarms and operator intervention.
4. Active safety instrumented functions not requiring manual activation.

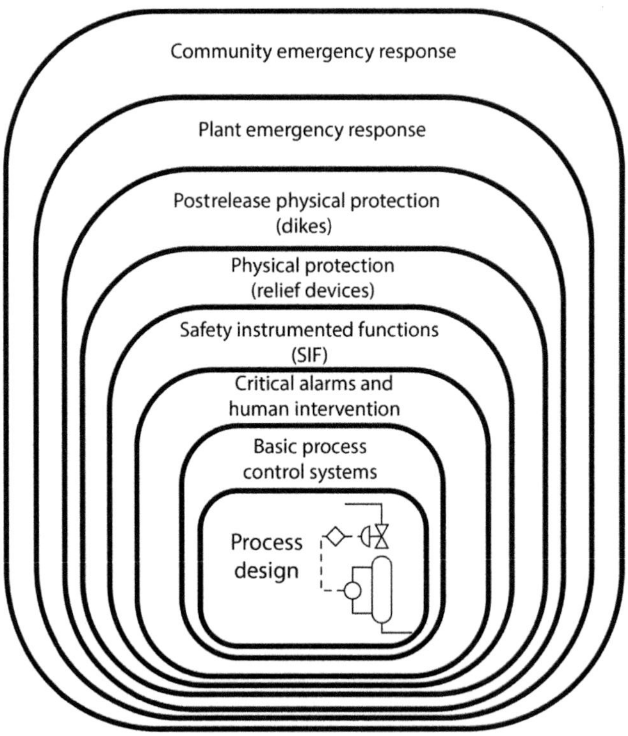


**FIG. 7.1** General view of the LOPA layers.

5. Physical protection such as relief valves.
6. Passive protection systems such as a bund or insulation, i.e. a secondary system that does not need to change its state when a demand is placed upon it. These are allowable for Environmental Cases and as mitigation in relevant scenarios such as pool fires.
7. Local (site) emergency response.
8. Community emergency response.

Fig. 7.2 shows the methodology of LOPA in the form of an event tree.

The final layers of the onion are the plant emergency response and the local disaster plan. In a gradually developing incident each layer is only required when the preceding one has failed to halt the development. To count in the analysis they must be truly independent of each other and, when they operate as intended, halt or mitigate the final unwanted outcome. It is essential to recognise that they do not alter the frequency of the final event.

One of the more difficult features of LOPA is the assessment of human presence and actions in a scenario, for example:

- The probability of a person being present at the hazard site.
- The probability that that person will be killed/injured.
- The human factor in the causation as initiating event and rectification mechanism as in reaction to alarms.

The overall sequence of a LOPA study is shown in Fig. 7.3

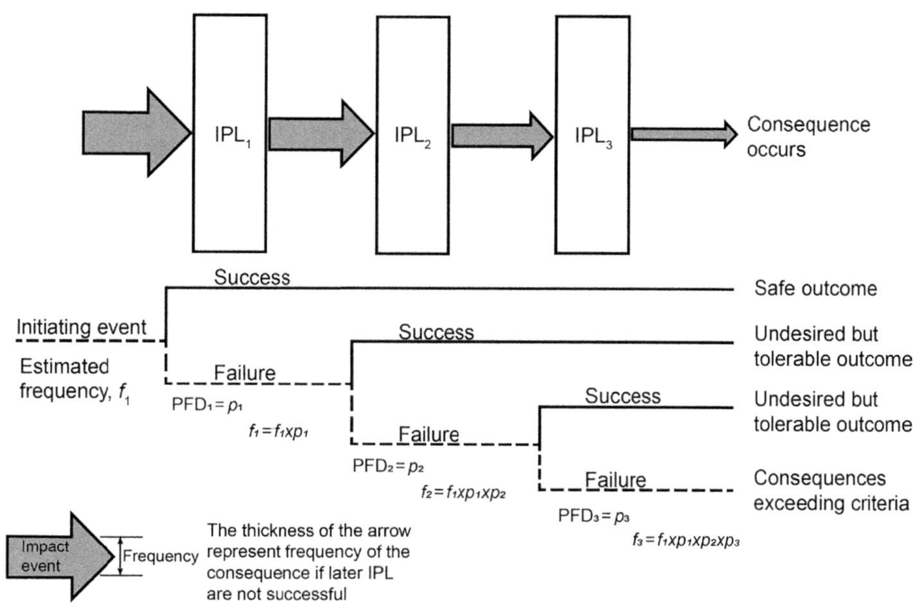


**FIG. 7.2** Methodology of LOPA in the form of an event tree.

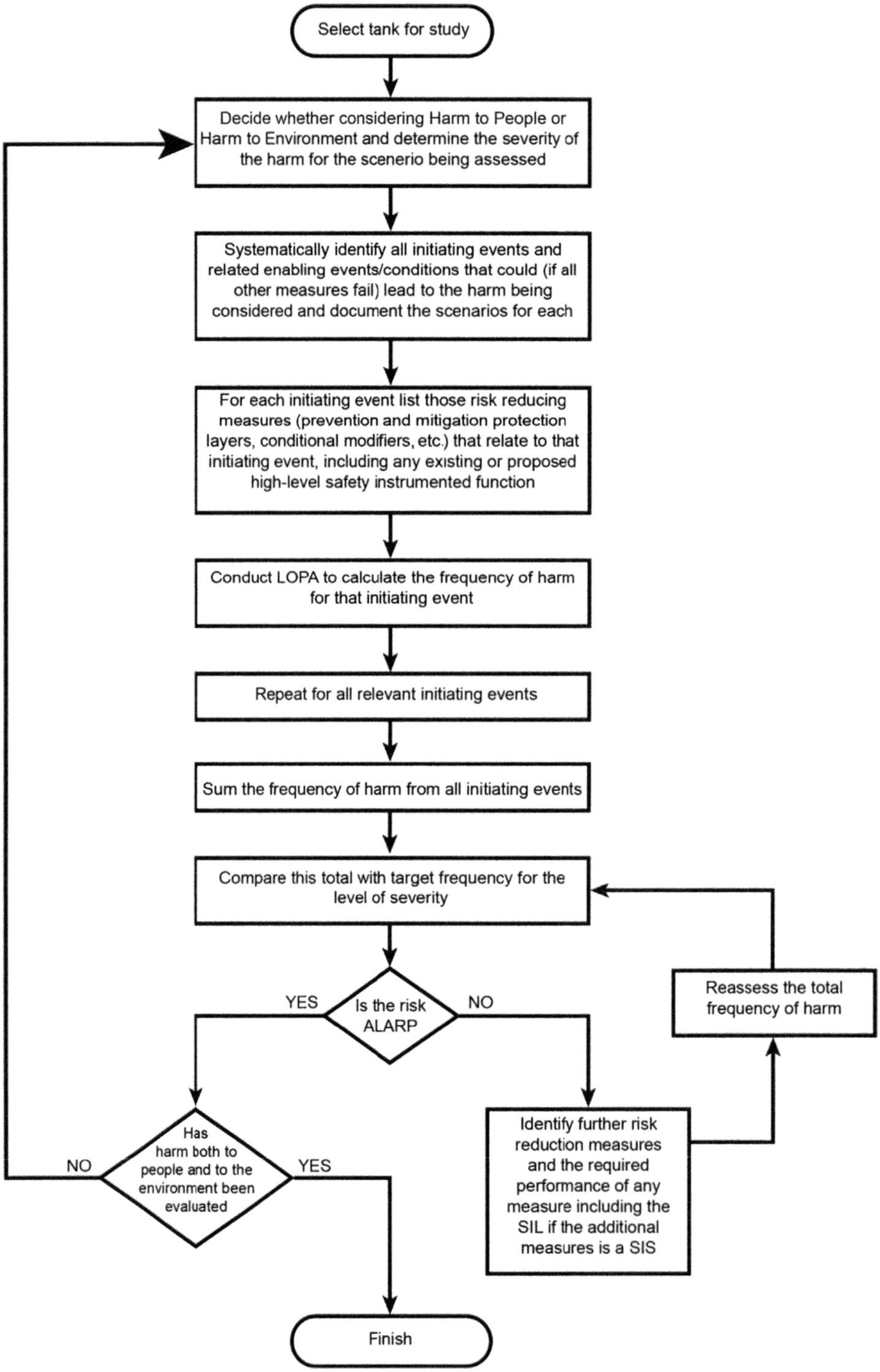


**FIG. 7.3** Flow chart for LOPA.

The starting point in a LOPA study is the identification of hazardous scenarios—major consequences that would result from a failure or deviation as the worst case if none of the preventative controls or mitigating features intervened. In most operating facilities, and certainly in major ones, there will be several significant scenarios. Earlier HAZOP and HAZID studies will identify many, though probably not all, of the possible scenarios. Each one will be examined separately. The first requirement is to identify the severity of the consequences, for example in terms of harm to persons, on or off site, or to the environment. Estimates of release rates and quantities will be needed to make possible a judgement of the worst outcome on personnel. Often proprietary software can be used to predict release rates, spread of vapour and radiation and blast effects from fires or explosions. Guidelines for toxic effects are available for common materials [5].

Next the 'target frequency' must be assigned for these consequences, using either company guidelines or those accepted by competent authorities such as the HSE. The following stages are perhaps best shown by reference to a LOPA worksheet that is in widespread use [6] (Table 7.1).

The scenario under examination is given an identification and description in the first row and the study team is recorded. Then the hazardous outcome—e.g. harm to persons—is given followed by the outcome—perhaps a fatality onsite—and the target frequency. The initiating causes must be identified along with their predicted frequencies in the right-hand column. These will normally be order of magnitude values. Ideally these come from user experience otherwise from equipment manufacturers or general databases. The team must be able to justify their choices. If human error is the cause of the primary initiating event it is best assessed by task analysis. Enabling conditions such as probability of ignition or probability of personnel being present are now recorded—the best estimates are entered and so are not necessarily rounded powers of 10. The spreadsheet automatically calculates the frequency of the unmitigated consequences.

In the next stage the LOPA onion layers are evaluated one by one to see if there is an applicable safeguard—an independent protection layer. For each one a description is inserted along with PFD (probability of failure on demand) for that item. Again, these will usually be order of magnitude values. The spreadsheet now calculates the total PFD, the final frequency of the consequences in the fully mitigated outcome and, by comparison with the previously estimated tolerable frequency for the unmitigated consequences, shows if the risk tolerance criteria has been met.

It is often the case that there are several initiating events for the same outcome and so each must be examined in this way. Then the cumulative frequency based on the sum from all these possibilities is used to determine whether the target has been met. If not, consideration must be given to closing the gap. With the clarity provided by the analysis of the individual cases it is usually possible to identify how this might best be done. If there are several ways then the most cost effective can be adopted.

On completion of each spreadsheet the entries should be reviewed and uncertainties and sensitivities clearly identified. There may well be uncertainty in the reliability data and of course the team should appreciate, which factors have the greatest effect on the

**Table 7.1** A typical spread sheet for use with LOPA.

| Scenario | PID/ item number | Scenario description | Reasoning/ justification | Study team | |
|---|---|---|---|---|---|
| 1 | | | | | |
| Consequence description/category | | | | | Frequency per year |
| Risk tolerance criteria | | | | | |
| Control system initiators | | | | | |
| Human factor initiators | | Description of possible error | Number of opportunities/ year | Probability of error per opportunity | |
| | | | | | |
| Other initiators | | | | | |
| **Enabling event or condition** | | | | | |
| Enabling events and conditional modifiers | | Probability of ignition (fire and explosion events) | | | |
| | | Probability ignition leads to explosion (if this is main event) | | | |
| | | Probability that personnel will be EXPOSED in affected area | | | |
| | | Others, e.g. time when risk is present | | | |
| **Frequency of unmitigated consequences** | | | | | |
| **Independent protection layers (with descriptions)** | | | | | PFD |
| BPCS actions with trip | | | | | |
| BPCS alarm and operator response independent of trip | | | | | |
| Pressure relief device—overpressure events | | | | | |
| Other safety related protection systems | | | | | |
| SIF A | | | | | |
| SIF B | | | | | |
| **Total probability of failure on demand for all IPLs** | | | | | |
| **Final frequency of consequences when all factors accounted** | | | | | |
| **Risk tolerance criteria met?** | | | | | |

outcome in terms of severity or frequency. For example, if an IPL has a very low PFD (this SIL meaning an expected reliability of 99.9%), then it must be maintained to this very high reliability throughout the life of the operating plant. This is a challenging task and explains the limited use of SIS of SIL3 or 4 outside of the aviation and nuclear industries.

Finally, of particular use in the United Kingdom, this spreadsheet extends to examine whether there is justification for further reduction in the predicted frequency to meet the principle of ALARP (see the list of Definitions and Abbreviations at the end of this chapter).

## 7.3 Resource requirements

The names and competences of the LOPA study team, for example:

- Trained Leader/facilitator
- Production operator(s) (for existing facility studies) or Project Engineer (for new facilities)
- Process Technology Specialist
- Process Control Specialist
- Production Engineer
- Maintenance
- Instrumentation design engineer
- Scribe (preferably recording and calculating software)

It may also be necessary to include someone who can assess consequential losses and cost and repair times in a meaningful manner.

Documentation will include

- Process Description
- Piping and Instrument Diagrams
- Operating Instructions
- Outcomes from previous studies (e.g. HAZOP)
- Lists of systems which are bypassed or in 'manual'
- Company procedures for LOPA studies, including corporate risk criteria for all consequences and reliability data

## 7.4 Timing

LOPA can follow on from a HAZOP study with a SIL assessment made for all the major events identified by the team. However, there are many advantages to carrying it out at an earlier hazard study (e.g. stage 2) as identification of the SIS needs at this stage will lead to more satisfactory and effective design work. LOPA must be done before the full assessment of the shutdown system. The method can also be used as part of a review of an existing plant but must be linked to an analysis of the hazards and their causes.

## 7.5 Advantages, disadvantages and uncertainties

### Advantages

- The use of LOPA highlights the required shutdown integrity at an early stage in the project study 2 (see Chapter 3).
- It can used as a quick screening tool to identify the need (or otherwise) for a simplex or more complex shutdown system.
- It provides a clearly structured tabulation and justification for the chosen integrity level and shows all the mitigating factors.
- It clarifies the requirements for the design, operation and maintenance of safety systems.
- Although the method is not primarily aimed at the identification of hazards it is capable of adding to those identified at the HAZOP study stage and to clarify their causes and outcomes.

### Disadvantages

- The quantification of the integrity of the independent protection layers/shutdown systems must be justified by a full and careful assessment.
- The assessment and quantification of the benefits of a protection layer may be subjective.
- The analysis of the higher SILs (2 and above) must consider common mode failures.
- The analysis of the higher SILs (2 and above) must consider human factors.
- Experience suggests that LOPA can reduce the rigour of the analysis.
- The use of a 5 × 5 risk matrix (based on decades) ignores the potential error bands of the data and analysis. This may require a more detailed risk assessment (Chapter 15) for potentially higher-risk incidents.

### Uncertainties

- It may be difficult to assess the true consequential losses for low-frequency events of large consequence.
- The uncertainty in estimates of frequency increases for low-frequency events and consideration must be given to a full risk assessment.
- Consequential losses and rebuilding costs are difficult to assess for unlikely scenarios.
- Some probabilities such as ignition probability are very difficult to quantify.

## 7.6 Example

This example is based upon the Buncefield fuel storage site explosion and fire in 2005. It is well documented in the HSE report ‘Safety and Environment Standards for Fuel Storage Sites’ [7]. A common reaction immediately after the event was surprise at the scale of the

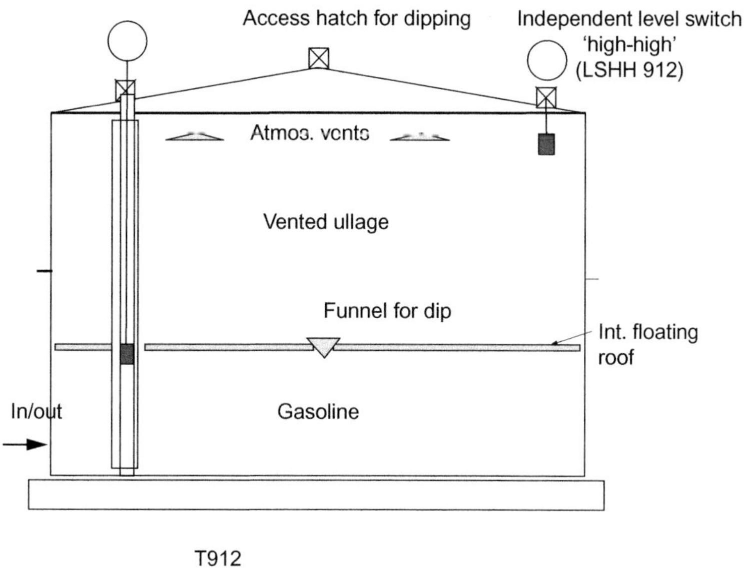


**FIG. 7.4** Simplified drawing of tank which overflowed at Buncefield T912.

explosion as the typical worst case envisaged for such sites was a pool fire. Subsequent review showed that an unconfined cloud explosion from such a spillage was not an unknown event but might have been anticipated. The investigations and reports include examination of the use of LOPA for both the preincident beliefs as well as from the basis of later understanding. The former is illustrated here with the worst case taken as a pool fire in the bund surrounding the tank (Fig. 7.4).

The event analysed is for an overflow from the largest fuel storage tank on the site, T912, operating within the normal range of fill rates facilitated by failure of the level indicator. The automatic tank gauge (ATG) provided a continuous readout of level to the control room panel and gave a high-level alarm. In addition there was an independent high-high level switch, LSHH912 designed to give automatic shut-off of the inlet flow with a further alarm. The chosen worst case is one fatality due to exposure of the patrol operator for which the selected risk tolerance criteria is 1.0E−5 events per year. For the size of spillage the ignition probability is taken as 0.12 (from Ref. [8]) with a 10% chance that the operator will be in the affected area and recognition that filling occurs for only 8% of the time. These combine to give a frequency for the unmitigated consequences as 9.6E−5. The only independent protection layer (IPL) is the high-level trip LSHH912 for which an assumed PFD of 0.1 is used.[a] So overall the frequency of consequences is the product of these two values, i.e. 9.6E−6 or just less than the target of 1.0E−5. Table 7.2 shows the record of this scenario examination:

[a]This is the lowest failure frequency acceptable within IEC16511 and must be justified.

**Table 7.2** Example of an analysis of Buncefield using LOPA.

| Scenario | PID/ item number | Scenario description | | Scale/severity of event is mitigated by: | Study team |
|---|---|---|---|---|---|
| 1 | T912 | Overflow at 500–900 $m^3$/h—pool fire | | Containment to minimise potential exposure of persons | Separate list |
| Consequence description/category | | Harm to persons | | | Frequency per year |
| Risk tolerance criteria | | Patrol operator exposed to pool fire | | 1 Fatality on site | 1.0E − 5 |
| Control system initiators | | T912 level indication system (ATG) fails | | BPCS control loop failure | 0.1 |
| Human factor initiators | | Description of possible error | Number of opportunities/ year | Probability of error per opportunity | |
| | | Initiator already entered | | | 1 |
| Other initiators | | Initiator already entered | | | 1 |
| **Enabling event or condition** | | | | | |
| Enabling events and conditional modifiers | | Probability of ignition (fire and explosion events) | Large overflow of gasoline ca. 100 kg/s | Value taken from standard tables | 0.12 |
| | | Probability ignition leads to explosion (if this is main event) | Pool fire only in this case | | 1 |
| | | Probability that personnel will be EXPOSED in affected area | Based on patrol operator's work pattern | 10% probability of exposure | 0.1 |
| | | Others, e.g. time when risk is present | Fraction of time when filling. 50 fills per year | 8% | 0.08 |
| **Frequency of unmitigated consequences** | | | | | 9.6E − 5 |
| **Independent protection layers (with descriptions)** | | | | | PFD |
| BPCS actions with trip | | Alarm from ATG not valid as it is the initiator | | ATG failed (common cause failure—NON INDEPENDENT) | 1 |
| BPCS alarm and operator response independent of trip | | LSHH912 trips feed inlet valve. PFD is based on test records and conforms with the lowest PFD allowed in IEC16511 | | BPCS shutdown is independent of initiating event | 0.1 |

**Table 7.2** Example of an analysis of Buncefield using LOPA—cont'd

| Scenario | PID/ item number | Scenario description | Scale/severity of event is mitigated by: | Study team |
|---|---|---|---|---|
| Pressure relief device—overpressure events | | Not relevant to case as an IPL | No relevant device | 1 |
| Other safety related protection systems | | None | | 1 |
| SIF A | | None | | 1 |
| SIF B | | None | | 1 |
| **Total probability of failure on demand for all IPLs** | | | 0.1 | |
| **Final frequency of consequences when all factors accounted** | | | 9.6E−6 | |
| **Risk tolerance criteria met?** | | | YES | |

However, the case is not complete at this point as there are other initiating causes of the same outcome (a pool fire) and each must be examined as the elements are unlikely to be the same for each one. The other causes are summarised below. They all have the same risk tolerance criteria of one fatality on site and hence a target frequency of 1.0E−5 but unmitigated frequencies and IPLs have some differences (Table 7.3).

The sum of the mitigated frequencies, 6.76E−4, shows that the target of 1.0E−5 is not met. This could be remedied by a SIS with a PFD of between 1E−01 and 1E−02, either a SIL1 system with a PFD of 1.48E−02 (Result 1E−05) or a SIL2 SIS which gives result 6.76E−06.

When it is agreed to make the changes indicated by this study, it is advisable to carry out a cost/benefit analysis for addition of further protection to assess if ALARP is met—that is to see if the cost of further reduction is grossly disproportionate to the benefit gained [9].

It should be noted that if the top event was taken the worst case of a vapour cloud explosion then the outcome of the LOPA study would be different. The exposed

**Table 7.3** Analysis of outcomes from example on Buncefield.

| Study | Initiating cause | Unmitigated frequency | PFD for all IPLs | Mitigated frequency | Target met? |
|---|---|---|---|---|---|
| 1 | T912 level indication fails due to BPCS control loop failure | 9.6E−5 | 0.1 | 9.6E−6 | YES |
| 2 | Operator fails to observe ATG (e.g. respond to high-level alarm) | 6.0E−3 | 0.1 | 6.0E−4 | NO |
| 3 | Wrong tank line-up | 6.0E−4 | 0.01 | 6.0E−6 | YES |
| 4 | Sending operator fails to stop transfer on request | 6.0E−3 | 0.01 | 6.0E−5 | NO |

population would have been greater and mitigation would be unlikely to contribute to reduction of the tolerable target frequency. The likelihood of persons' exposure would have been greater and the probability of an ignition leading to an explosion would be taken into account. Using LOPA reveals that added protection would be needed with higher integrity requirements for protection systems.

## Definitions and abbreviations

Commonly used terms in LOPA include:

- BPCS. Basic Process Control System. The control system that operates to maintain the required process parameters.
- IPL. Independent Protection Layer. A totally independent protective system, which is not replicated elsewhere. It must act independent of any other protection and be unaffected by the initiating event.
- SIS. Safety Instrumented System. An instrumented system, which detects possible process parameters, which could lead to an incident and then takes executive, preventative, action through a shut-off valve or equivalent. It must be independent of any other protection and the initiating event and usually automatically takes the facility to a safe state.
- SIF. Safety Instrumented Function. The required complete action of a SIS when activated. This means the function of the SIS loop from detection through to the action of the final element, which interrupts to prevent the hazard.
- IE. Initiating Event or Cause. An event or combination of circumstances, which might create a potentially hazardous situation.
- Demand. The condition, which requires the corrective action of a protective system.
- CCF. Common Cause Failure (mode) The failure of one or more systems through a common cause. [This could be a fire, overheating, corrosion—see Risk Assessment (Chapter 15).]
- CMF. Common Mode Failure. The failure of one or more systems in the same manner.
- SIL. Safety Integrity Level. This is a measure of the performance required from a safety instrumented system to maintain or achieve the safe state A SIL rating may be from 1 to 4. A SIL1 system must have a probability of failure on demand of between 1E − 1 and 1E − 2. SIL4 systems require redundant and diverse systems and are rarely found outside the nuclear and aviation industries.
- Redundant. A system that has at least two channels of detection or action. Higher-grade systems may include a voting system based on 2 out of 3 channels.
- Diverse. A system that has parallel channels, which detect different basic properties of a process, which are related to each other directly. For example, temperature and pressure may be a diverse system for the detection of a pressure deviation.
- PFD. Probability of failure on demand. The probability that a system will fail to act as designed when a demand is placed upon it. Note however that Chemical Engineers have for many years also used this acronym for Process Flow Diagram.

- Scenario. An unplanned sequence of events leading to an incident with undesired consequences.
- ALARP. As low as reasonably practicable. Defined in HSE literature [9].

## References

[1] CCPS, Layer of Protection Analysis: Simplified Process Risk Assessment, 2001.

[2] CCPS, Guidelines for Initiating Events and Independent Protection Layers in Layer of Protection Analysis, 2015.

[3] CCPS, Guidelines for Enabling Conditions and Conditional Modifiers in Layer of Protection Analysis, 2014.

[4] R.T. Gowland, How to LOPA, Chem. Eng. (May) (2016).

[5] ERPG values provided by the American Industrial Hygiene Association via their website.

[6] Courtesy of R.T. Gowland. As demonstrated and used at the IChemE LOPA training courses.

[7] HSE, Safety and environmental standards for fuel storage sites, Process Safety Leadership Group, (Final report). 2009.

[8] OGP, (Risk Assessment Directory Report No. 434-6). Ignition probabilities, 2019.

[9] Reducing Risk Protecting People (R2P2), 2001, HSE Publications.

# 8 Relative ranking

Chapter outline



## 8.1 Definition

Relative ranking is a structured method of analysis leading to a numerical grading of the main hazards associated with each section of a plant or process. These grades may be used in a number of ways, for example to identify low hazard options or to select sections that require further hazard analysis. The hazards of concern usually include fire, explosion, airborne releases of toxic substances and in some cases releases of materials dangerous for the environment.

## 8.2 Description

The most widely used ranking method for fire and explosion is the Dow Fire and Explosion Index (FEI), augmented by the Dow Chemical Exposure Index (CEI) for airborne releases of toxic substances and both are described here. Detailed guides are available for both the FEI [1] and the CEI [2] and further accounts appear in standard references [3, 4], which also mention some other ranking methods. A specialised index has been developed and tested for offshore oil and gas installations [5] and the basic methods can be extended if required but these are not covered here.

The Dow FEI, first published in 1964, is now in its seventh edition. It is widely used, both within the Dow Chemical Company and by others, primarily to give a relative value for the potential of process unit losses due to fire and explosion hazards. Dow has recently extended the method into a software version, which is used internally.


A Guide to Hazard Identification Methods. https://doi.org/10.1016/B978-0-12-819543-7.00008-2

The guide provides the necessary detail, formulae, graphs and methodology for an experienced engineer to carry out an assessment. The initial output from the FEI is a value for the index for that unit of the plant. The index by itself may be of direct use, for example the Dow Chemical Company requires that all facilities are required to carry out a Process Hazard Analysis (PHA) on a 3-year cycle and whenever the leadership of a facility changes or a significant new project or development is planned. When the FEI within the PHA process reveals an index of 128 or higher, a Layer of Protection Analysis (LOPA) is required. In addition, technology new to the company always requires a HAZOP. The method may be taken further to get estimates of the realistic maximum losses for the units, expressed in terms of an Actual Maximum Probable Property Damage (Actual MPPD), Maximum Probable Days Outage (MPDO) and Business Interruption (BI). In estimating these values, an area of exposure and a damage factor are also obtained. These later estimates apply well for fire cases, but if an explosion is a required case, Explosion analysis using PHAST or TNO Multi Energy methods are often applied.

The first step in applying the FEI to a plant is to select the key units—known as pertinent process units—within the plant. The selection is based on the individual unit operations such as unloading system, storage, tank, reactor, distillation column, etc. type of material within the unit, the quantity of material, the process conditions, the cost of equipment, the presence of items critical to operation and any history of problems. A process seldom requires a detailed analysis for more than 3 or 4 units. A material factor (MF) is derived for the materials contained in each unit using standard NFPA flammability and instability rankings for the dominant material that is present. The guide provides an extensive list of these and a means of estimating the MF for unlisted materials or for mixtures.

Two penalty factors, one for general process hazards (F1) and the other for special process hazards (F2), are then established. The general process hazards cover exothermic and endothermic reactions, material handling and transfer, enclosed process units, access, and drainage and spillage control. Special process hazards are based on the toxicity of the materials being processed, low- and high-pressure working, operation near a flammable range, dusts, low-temperature operation, the quantity of material in process, corrosion and erosion, leakage and a few other unusual aspects such as the use of fragile process containing equipment such as glass, hoses, expansion joints. The analyst determines whether or not a feature applies and, if so, chooses an appropriate penalty factor following the guidance provided or, for pressure and quantity, using the given empirical correlations. The guide includes examples showing how these factors should be estimated. The penalties are then totalled to give F1 and F2 and the product of these with the MF gives the value for the FEI. Based on the extensive experience of using this method the unit is then classified in one of five bands, from light to severe. A useful effect of using the index during projects is that modifications can be assessed from an Inherent Safety aspect, since a reduction in index usually accompanies an Inherently Safer design.

In a full analysis, the next step is to select and assign values to loss control credit factors. These categories are process control, material isolation and fire protection. Features such as emergency shutdown, operating procedures, other hazard analyses, interlocks, fire-cladding, etc. earn credit factors, which are usually in the range of 0.90–1.0 with a few below 0.9. An equation is given to derive the overall Loss Control Factor from the relevant credit factors. Apart from their use in the calculation of the Actual MPPD, the derivation of these factors provides a valuable record of the safety features within system. These must be maintained during the plant lifetime if target safety levels are to be achieved. The final step of the analysis is now straightforward. An area of exposure and a damage factor are derived from the MF, F1 and F2. These are used in combination with the replacement value of equipment and materials in that area to calculate the remaining factors (Base MPPD, Actual MPPD, MPDO and BI). The relationships used in the final calculations rely upon historical incident data and some judgement is necessary in the application to each unit. The radius of damage is useful in determining separation of key process units to avoid 'domino' effects.

The Chemical Exposure Index (CEI) provides a ranking method for the acute toxicity health hazard potential from hypothetical chemical releases on a plant. As with the FEI, the documentation is available through the AIChE [2].

Initially various possible release scenarios are considered using experience-based rules as to likely ruptures and failures. An airborne quantity is calculated (actually a vapour generation rate) based on the type of release. Possibilities include gases, flashing liquids and liquids forming pools, including partial flashing if appropriate. This is combined with an emergency exposure level (ERPG-2 [6] or equivalent), which, over 1 hour, should not cause irreversible health effects or impair the ability to take protective action. The value of the CEI is then obtained by formula as are hazard distances—an indication of the distance within which the ERPG is exceeded.

The latest issue of the AIHA Guide to ERPG values should always be consulted. If no value is available, the CEI Guide gives advice on how to proceed. Detailed examples illustrate the main release scenarios and a containment and mitigation checklist is provided to help the review process. The Dow Chemical Company requires units having a CEI above a threshold value of 200 to be subjected to a further risk review. Whilst the CEI does not provide an absolute measure of risk or precise values of hazard distances it is a well-based method of ranking different scenarios across a wide range of processes, materials and possible incidents. It can be used for screening purposes, to identify the main sources of toxic hazards and to help with the mitigation of them, to compare different options or to show the possible effects of a modification. The only difficulty in its use occurs when there is no published ERPG. When ERPG2 concentrations are predicted to exceed fence-line distance, deeper study is indicated (e.g. PHAST) and LOPA.

## 8.3 Resource requirements

*Manpower*: The method can be used by an individual, although there are advantages in having two or more involved, and it is always advisable to have the results reviewed by an independent person. It is possible for an untrained, but experienced, engineer to apply the methods although training or initially working with an experienced assessor is helpful.

When the data are available and a site visit has been made for existing plant, an experienced user can normally complete the assessment on a unit in 1–2 hours.

*Data*: The analyst will need a site plan and a minimum of a block diagram of the plant showing the main items of hardware, including the actual or intended layout; overall design and operating data; a list of all the materials, quantities, full material and safety data sheets; the appropriate guide and recording forms.

## 8.4 Timing

For a new project the technique would always be used at an early stage, either when alternative routes are being explored, when different scales of operation and the possible process methods are being evaluated and, after a final decision, before the detailed design work in undertaken.

For existing plant it may be used to review the operation, especially after a period of operation, or whenever a change of process or modification of the plant is proposed.

## 8.5 Advantages, disadvantages and uncertainties

### Advantages

- Relatively simple with limited need for plant and process information.
- Can be applied at an early stage in a project before a detailed P&ID is available.
- Indicates the main areas with hazard potential.
- The methodology can be adapted to suit company needs.
- The method supports the drive to inherently safer design, since it is clear that a process unit with a lower index is inherently safer since it prioritises the principles of moderation, reduction, etc.

### Disadvantages

- The methods are limited to the hazards of fire, explosion and chemical exposure.
- Skilled interpretation is needed. In Dow, this is carried out by the Process Safety Engineer function.
- Detailed features of the design, which may lead to hazardous events are not highlighted.
- The methods usually need to be used in conjunction with other hazard identification and assessment methods.
- The method was developed for use with mainstream chemical activities. It cannot be used directly in some others such as the off-shore oil and gas industry.
- It does not explicitly cover environmental problems.
- Little account is taken of human factors.

### Uncertainties

- Although the starting point (the MF) is precise, many of the subsequent factors involve a judgement by the analyst. Differences in the final result of ±10% can occur. It is also possible to have differences in the sectioning of the plant.
- The levels for action are based upon the use of the technique on many plants over many years. Where this basis is not available or not applicable it is difficult to interpret the numerical values.

## 8.6 Applications for relative ranking methods

The calculations from the Dow FEI can be used

- as a guide to the selection of fire protection methods;
- to focus attention on key areas or items of equipment;
- for the comparison of alternative process routes;
- to encourage use of the inherent safety;
- to evaluate the effects of modification;
- to act as a filter before the application of more detailed hazard study methods;
- to follow trends within a site over time;
- to help in planning layouts and separation distances.

Indices are particularly valuable in the early stages of a project due to their simplicity and speed of use and also as a screening tool for the application of other hazard review methods in the case of both new and existing plants. For multisite companies they can provide a means of standardisation across varied plants, processes, management and regulatory regimes.

The CEI can be used in many of the above ways, particularly in directing attention to the most significant scenarios and helping to ameliorate these and as an aid to effective Emergency Response Planning.

## 8.7 Example of the Dow FEI [1]

The application of the Dow Fire and Explosion Index to a Butadiene unloading operation.

The facility is an outdoor unloading station where railcars of 20,000 kg of 1,3-butadiene are unloaded into a storage tank. The connection to the inlet of the storage tank is via an unloading arm, which replaced the armoured hose used previously. The are two connections

**(1)** to the bottom outlet of the railcar
**(2)** to the top vapour space of the railcar

The first of these is to allow the butadiene to flow to the storage tank. The second is connected to a compressor which compresses butadiene vapour from the top of the storage tank to provide the driving pressure for the unloading operation. The overall operation is described as 'in process' rather than 'in storage' because the case involves several operating steps from the positioning of the railcar, connection, transfers, disconnection and isolation.

The operation is outdoors, with good access and a modern drainage system to a containment pond.

In the sections below each entry into the FEI worksheet in Table 8.1 is discussed and explained.

**Material factor**: The material factor for Butadiene is 24 (from the NFPA ratings in the appendix of the manual).

1. **General Process Hazards**: Exothermic or Endothermic Reactions
   No reactions take place, so no penalty is applied.
   **Material Handling and Transfer**: This operation clearly falls into this category.
   A penalty of 0.85 is applied (Flammability rating 4).
   **Enclosed or Indoor Process Units:** No penalty is applied for an outdoor operation.
   **Access**: In this example, the access for emergency response is good, so no penalty is applied.
   **Drainage and Spill Control.** This area is well drained to a containment system. In view of the fact that Butadiene has a high vapour pressure, drainage is not deemed to be a major factor. No penalty is applied.
   Applying the penalties with the base factor of 1.00 results in a **General Process Hazard (F1) of 1.85.**
2. **Special Process Hazards**:
   A. **Toxic Materials**: Butadiene has a NFPA toxicity rating of 2, so a penalty of 0.4 applies.
   B. **Subatmospheric Pressure**: This question relates to subatmospheric pressure, which could cause air to be drawn into the system. The operating pressure is in excess of 300 kPa, so no penalty applies.
   C. **Operation In or Near the Flammable Range**: This operation is significantly above the upper flammable limit. No penalty applies.
   D. **Dust Explosion**: This is not relevant. No penalty.
   E. **Pressure:** The operating pressure is 300 kPa, the relief system pressure is 450 kPa. The system calculated penalty is 0.12 when the multiplier for LPG type materials is applied.
   F. **Low-Temperature Operation**: No penalty is applied.
   G. **Quantity of Flammable Material—IN PROCESS**: is 20,000 kg and the Heat of Combustion is 10,700 kcal/kg. This produces a penalty of 1.38.
   H. **Corrosions and Erosion**: The material of construction is stainless steel. No corrosion is experienced with this material in butadiene service. No penalty is applied.

**Table 8.1** FEI worksheet [1] for butadiene unloading example.

| AREA / COUNTRY<br>EUROPE | DIVISION<br>LATEX | LOCATION<br>MANCHESTER | DATE<br>10/3/2019 |
|---|---|---|---|
| SITE<br>Manchester | MANUFACTURING UNIT<br>Plant 1 | PROCESS UNIT<br>Butadiene Unloading | |
| PREPARED BY:<br>Snooks | APPROVED BY: (Superintendent)<br>Snodgrass | BUILDING | |
| REVIEWED BY: (Management)<br>Sparks | REVIEWED BY: (Technology Center)<br>McGregor | REVIEWED BY: (Safety & Loss Prevention)<br>Smith | |
| MATERIALS IN PROCESS UNIT<br>Butadiene liquid and Vapour | | | |
| STATE OF OPERATION<br>__ DESIGN __ START UP _x_ NORMAL OPERATION __ SHUTDOWN | | BASIC MATERIAL(S) FOR MATERIAL FACTOR<br>1,3-Butadiene | |

| MATERIAL FACTOR (See Table 1 or Appendices A or B) Note requirements when unit temperature over 140 °F (60 °C) | | 24 |
|---|---|---|
| **1. General Process Hazards** | **Penalty Factor Range** | **Penalty Factor Used(1)** |
| **Base Factor** ........ | 1.00 | 1.00 |
| A. Exothermic Chemical Reactions | 0.30 to 1.25 | 0 |
| B. Endothermic Processes | 0.20 to 0.40 | 0 |
| C. Material Handling and Transfer | 0.25 to 1.05 | 0.85 |
| D. Enclosed or Indoor Process Units | 0.25 to 0.90 | 0 |
| E. Access | 0.20 to 0.35 | 0 |
| F. Drainage and Spill Control ________ gal or cu.m. | 0.25 to 0.50 | 0 |
| **General Process Hazards Factor ($F_1$)** ........ | | 1.85 |
| **2. Special Process Hazards** | | |
| **Base Factor** ........ | 1.00 | 1.00 |
| A. Toxic Material(s) | 0.20 to 0.80 | 0.4 |
| B. Sub-Atmospheric Pressure (< 500 mm Hg) | 0.50 | 0 |
| C. Operation In or Near Flammable Range __ Inerted __ Not Inerted | | 0 |
| 1. Tank Farms Storage Flammable Liquids | 0.50 | 0 |
| 2. Process Upset or Purge Failure | 0.30 | 0 |
| 3. Always in Flammable Range | 0.80 | 0 |
| D. Dust Explosion (See Table 3) | 0.25 to 2.00 | 0 |
| E. Pressure (See Figure 2) Operating Pressure __300kPa__ psig or kPa gauge<br>Relief Setting __450kPa__ psig or kPa gauge | | 0.116 |
| F. Low Temperature | 0.20 to 0.30 | 0 |
| G. Quantity of Flammable/Unstable Material: Quantity 20000kg lb or kg<br>$H_C$ = _10,700 kC/kg__BTU/lb or kcal/kg | | |
| 1. Liquids or Gases in Process (See Figure 3) | | 1.38 |
| 2. Liquids or Gases in Storage (See Figure 4) | | |
| 3. Combustible Solids in Storage, Dust in Process (See Figure 5) | | |
| H. Corrosion and Erosion | 0.10 to 0.75 | 0 |
| I. Leakage – Joints and Packing | 0.10 to 1.50 | 0.1 |
| J. Use of Fired Equipment (See Figure 6) | | 0 |
| K. Hot Oil Heat Exchange System (See Table 5) | 0.15 to 1.15 | 0 |
| L. Rotating Equipment | 0.50 | 0 |
| **Special Process Hazards Factor ($F_2$)** ........ | | 3.00 |
| **Process Unit Hazards Factor ($F_1$ x $F_2$) = $F_3$** ........ | | 5.54 |
| **Fire and Explosion Index ($F_3$ x MF = F&EI)** ........ | | 133 |

I. **Leakage, Joints and Packing:** There are seals on the compressor and on the loading arm connections. A penalty of 0.1 is applied.
J. **Use of Fired Equipment**: This refers to operations like fire burners and furnaces. In this example, the nearest such operation is at least 300 m distant. No penalty is applied.
K. **Hot Oil Heat Exchange System**: There is no heating in this operation. No penalty is applied.

3. **Special Process Hazard (F2) calculates at 3.00.**
4. **The Fire and Explosion Index calculation is carried out. F1 × F2 × Material Factor yielding the result is 133.** In Dow Chemical risk range ranking, this would result in study such as their form of HAZOP and Layer of Protection Analysis.

## 8.8 Conclusion

This example demonstrates the ease with which an FEI analysis can be made for a routine, although potentially hazardous, operation can be assessed. The result, using the Dow Chemical risk range ranking, is that a further, more detailed, method of hazard identification and assessment should be applied. It also indicates that it can be used both for a new process operation as well as for existing plant.

## References

[1] Dow Chemical Company, Fire & Explosion Hazard Classification Guide, seventh ed., American Institute of Chemical Engineers (AIChE), New York, USA, 1994 (rev 1997).

[2] Dow Chemical Company, Chemical Exposure Index Guide, second ed., American Institute of Chemical Engineers (AIChE), New York, USA, 1993 (rev 1997).

[3] S. Mannan (Ed.), Lees' Loss Prevention in the Process Industries: Hazard Identification, Assessment and Control, fourth ed., ISBN: 0123971896, 2012.

[4] Guidelines for Hazard Evaluation Procedures, third ed., CCPS, ISBN: 978-0-471-97815-2, 2008.

[5] F.K. Crawley, M.M. Grant, Concept risk assessment of offshore hydrocarbon production installations, Trans. IChemE 75 (Part B) (1997) 157.

[6] Emergency Response Planning Guideline as issued by the American Industrial Hygiene Association. The 2016 list covers 149 compounds. It is periodically updated.

## Further reading

[7] F.I. Khan, T. Husain, S.A. Abbasi, Safety weighted hazard index, Trans. IChemE 79 (Part B) (2001) 65.

# 9 The risk analysis screening tool (RAST)*

Chapter outline



## 9.1 Definition

RAST is a method for establishing an initial list of risk scenarios derived from an extensive data base adapted for the specific unit in a hazards and risk analysis. It links directly to further LOPA studies.

## 9.2 Description

RAST is a collection of process safety and risk analysis screening tools to aid evaluation teams when performing Hazard Identification and Risk Analysis (HIRA) studies by

*The Risk Analysis Screening Tool (RAST) and its associated Chemical Hazard Engineering Fundamentals (CHEF) documentation [1] have been developed through the collaborative efforts of volunteers from member companies of the Center for Chemical Process Safety (CCPS) and the European Process Safety Centre (EPSC). Special appreciation is extended to The Dow Chemical Company for donating RAST/CHEF for global use and both Dow and the CCPS for providing the resources to help modify the software and documentation.

A Guide to Hazard Identification Methods. https://doi.org/10.1016/B978-0-12-819543-7.00009-4

- What are the hazards?
- What can go wrong?
- How bad can it be?
- How often might it happen?
- Is the risk tolerable?

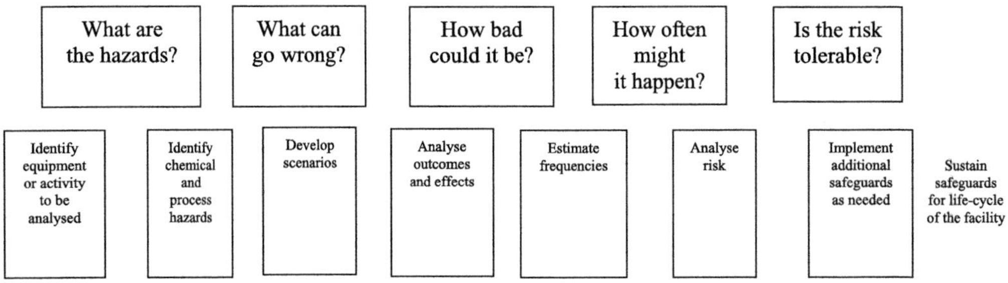


**FIG. 9.1** Hazard Identification and Risk Analysis work flow.

providing consistency among analysis teams while reinforcing company protocol and criteria. It utilises simplified and often empirical methods in quantifying hazards, consequences and risk. RAST allows users to enter the chemicals, reactivity data, equipment type, operating conditions (e.g. pressures, temperatures) and facility layout. RAST then uses these inputs to provide an initial list of risk scenarios based on its internal data base or HAZID (Chapter 5), Checklists (Chapter 10) and What if? (Chapter 11), and then estimates the 'worst' consequences. Once the Hazard Evaluation is completed, additional study teams can select specific scenarios for a Layer of Protection Analysis (LOPA; Chapter 7) working within the RAST software.

The timing of the study is quite critical as there may be significant cost implications if issues are identified on a new project after it is sanctioned.

RAST is based on a suggested HIRA work process (Fig. 9.1) to answer basic questions.

RAST and the accompanying Chemical Hazards Engineering Fundamentals (CHEF) materials are based on performing HIRA tasks in a specific order. This is based on an overall work flow such that results of a specific estimate (such as a source model) are available as input for the subsequent task (such as vapour dispersion). RAST uses a minimal information set initially with more information added over time to improve the analysis. The RAST User's Manual [1] provides the necessary detail for an experienced study team to perform an evaluation. The CHEF Manual [1] provides formulae, graphs and documentation of the methodology.

RAST does not answer all the questions posed during a thorough hazards evaluation and, specifically, is not a stand-alone programme; nor should it be used by one person to make any risk-based decisions.

## Using RAST

As is illustrated in Fig. 9.1, the overall work flow process for performing a HIRA within RAST is:

**(1) Identify the equipment or activity**: RAST looks at a specific activity involving a chosen item of equipment and its contents, for example the operation of a storage tank, a reactor or a piping network. Inputs are chemical data, equipment design information, operating conditions, and plant layout.

**(2) Identify chemical and process hazards**: This step identifies the hazards, the inherent chemical or physical characteristics that have the potential for causing damage to people, property or the environment. RAST considers both chemical and operational-related hazards.

Chemical hazards include flammability, toxicity, corrosivity and reactivity (stored chemical energy; or instability). The data for more than 250 chemicals are provided within the programme and users may enter properties for additional chemicals as needed in the HIRA study.

Operational hazards include stored pressure-volume energy, high or low temperature (potential for thermal burns) and, to some extent, electrical conductivity (potential for static charge accumulation and discharge).

Full details of the data requirements and how they are used are provided in the user manuals [1].

**(3) Development of scenarios (Table 9.1)**: In this step the RAST software suggests some possible unplanned events or incident sequences that result in a loss event—a set of scenarios drawn from the RAST library of common scenarios. For each one the associated impacts, including the success or failure of safeguards involved in the incident sequence, are given. Each RAST scenario contains one unique loss event and is linked to a specific source term that allows RAST to perform a simple consequence analysis. The suggested list of scenarios provided by RAST is not intended to represent all scenarios needed for an effective HIRA study, but it is an excellent starting point that

**Table 9.1** Some elements for scenarios included in RAST.

| Partial list of initiating events | Partial list of loss events | Partial list of incident outcome |
|---|---|---|
| Human error | Small hole leak | Flash fire |
| Mechanical failure | Medium hole leak | Building explosion |
| Regulator failure | Full bore leak | Vapour cloud explosion |
| Pump seal failure | Overfill release | Fireball or BLEVE |
| Heat exchanger tube failure | Vapour release—fire | Toxic release |
| Hose failure | Vapour release—reaction | Toxic infiltration |
| Loss of agitation | Equipment rupture | Environmental damage |
| Utility failure | Equipment damage | Business loss |

the evaluation team may build upon (and can add to if particular incidents known to the team apply). The pattern for a RAST scenario development is illustrated in Fig. 9.2.

**(4) Consequence analysis**: This step in RAST uses various source and effect models from CCPS [1] and other literature sources. The release loss events are categorised as related to hole size (vapour, liquid or two-phase), material balance (such as overfill), heat balance (such as vaporisation resulting from fire exposure), rupture (instantaneous release) or equipment damage. If the release is liquid or two-phase, vapour rate is estimated from simple flashing, aerosol evaporation and pool evaporation models. Fig. 9.3 shows the generic event tree used within RAST to determine incident outcomes.

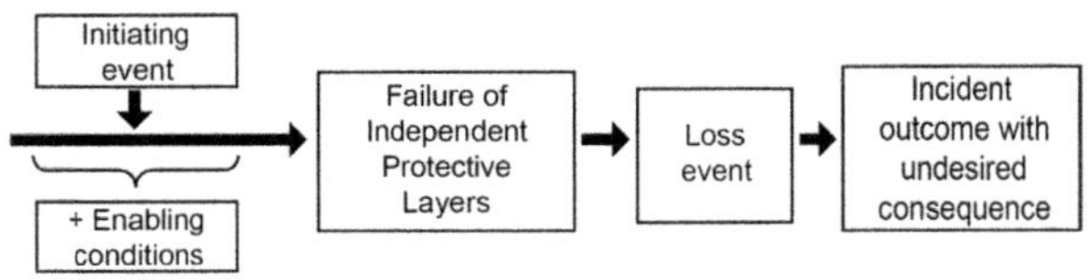


**FIG. 9.2** Model for scenario development in RAST.

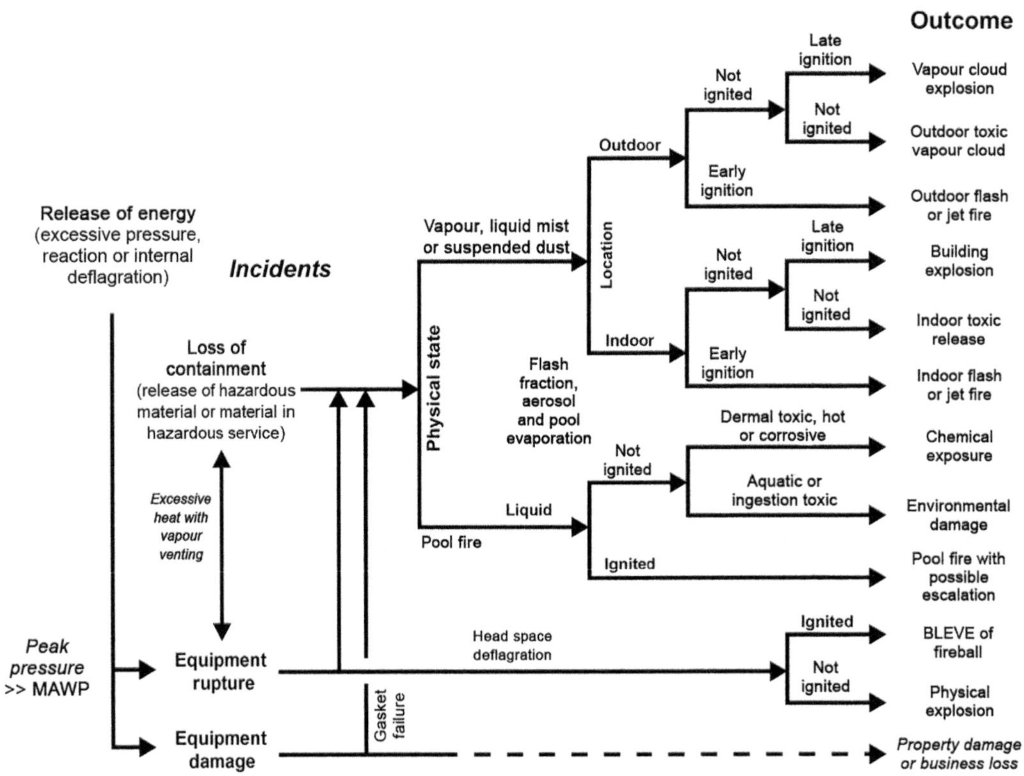


**FIG. 9.3** Generic incident outcome event tree used in RAST.

RAST estimates a single 'worst' consequence severity for each incident outcome. It offers alternative means in assessment of human harm and also allows the team to use their own rankings of severity level to accord with the company policies. Full details of these options are given in the manual [1].

**(5) Scenario frequency**: This step in RAST allows users to evaluate an order-of-magnitude frequency level and is based on the independence between the initiating events, the enabling conditions/conditional modifiers and the protective layers. This means that issues, which are on the risk parameter margins must be treated as absolute and should subject to a full QRA. Pre-populated tables store industry-generic initiating event frequencies, enabling condition or conditional modifier probabilities (such as probability of ignition), and probability of failure upon demand for Independent Protection Layers (IPL). These tables and correlation coefficients can be modified, as needed, to match the company's specific tolerable frequency values.

**(6) Risk analysis**: A consequence/frequency risk matrix is now used to categorise the risk for each scenario. The default risk matrix provided by the RAST software has 6 ranges for frequency and 5 classes of severity (consequence impact). This matrix is based on generic, industry based, risk assessment criteria. Using the simplified risk equation (i.e. risk equals the frequency times the severity), risk is rated in these categories:

- acceptable risk
- tolerable risk onsite
- tolerable risk offsite
- unacceptable risk.

These may create a grey zone at the margins of the matrix and therefore must not be treated as absolute values. Each company must develop its own risk criteria and modify the RAST matrix default parameters as needed. The risk criteria should include the following:

- Injury
- Fatalities on and off site, single or multiple and immediate or delayed.
- Capital loss
- Consequential loss, including all aspects of business
- Environment harm
- Adverse public reaction
- See also Risk Assessment (Chapter 15)

**(7) Layer of Protection Analysis (LOPA)**. This is an optional step in the RAST sequence and allows a designated LOPA team to use RASTs' integrated LOPA workbook to evaluate the protection layers that are needed for selected 'worst' scenarios, i.e. those with the greatest risk before adding engineering and administrative controls. RAST utilises a generic bow-tie relationship to determine which scenarios may require the greatest number of protective layers for each loss event (Fig. 9.4).

**RAST documentation**: RAST maintains and updates datasets of chemicals, suggested scenarios, consequence analysis results, and LOPA results (if applicable) for each equipment item evaluated. Users may add to or amend their existing dataset. When the study is

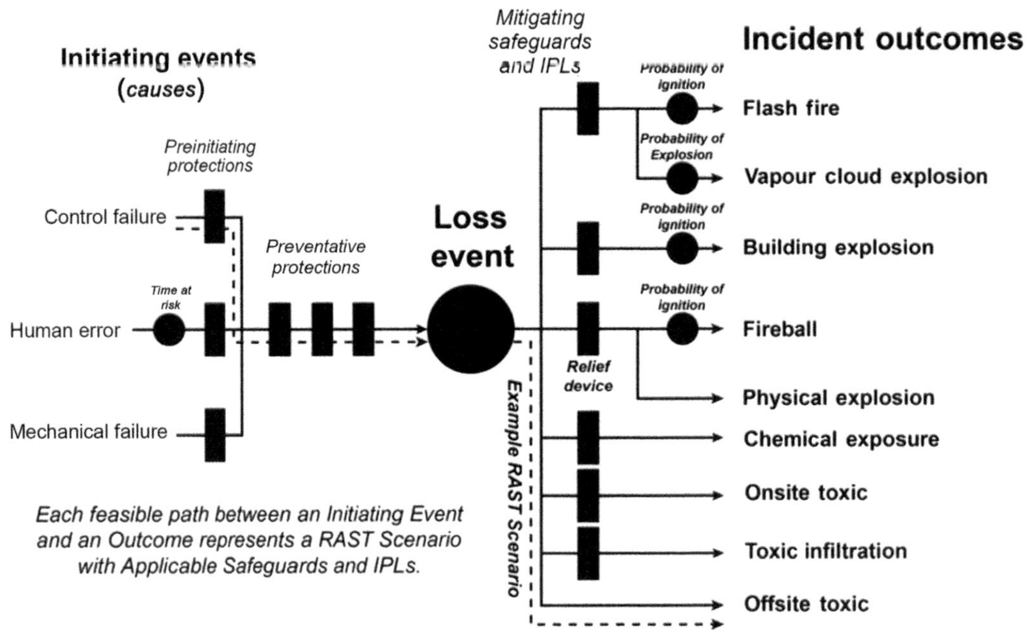


**FIG. 9.4** Generic bow-tie diagram used in RAST to determine 'worst' scenarios. See also Chapter 18, Bow-tie diagrams.

completed, all the assumptions and parameters used in the scenario evaluations have been documented. Not only does this documentation provide future hazards evaluation teams with an excellent foundation from which to start their revalidation efforts, this documentation can be used when process or equipment changes due to improvements or modifications.

**RAST training materials**: There are two manuals available for RAST [1]. The Chemical Hazard Engineering Fundamentals (CHEF) manual is intended for newer engineers or as a refresher for experienced personnel. It describes mathematical assumptions and the methodology for performing a Hazard Identification and Risk Analysis (HIRA) study. The Risk Analysis Screening Tool (RAST) user's manual focuses on how to utilise the software in helping HIRA study teams to improve productivity, consistency, and quality of the studies. Additional resources are available through the CCPS to provide guidance on updating the default values in the downloadable version of RAST so that each company can tailor the software to meet their risk tolerance criteria.

## 9.3 Resources

A study team should consist of a cross-section of personnel and would typically include:

- Study Leader/Facilitator
- Scribe and software recorder

- The team members participants bringing knowledge of equipment design, process design, operational experience, or maintenance experience to the study team. There are many similarities to the requirements for a HAZOP study team (see Chapter 4)
- An experienced RAST User—ensures appropriate inputs are entered into RAST to meet study objectives. Presents reports and results to the team to address questions and support decisions.

A RAST Technical Administrator (who may or may not be part of the study team) ensures that the 'administrative' parameters used in RAST-related efforts of the study team are consistent with company's protocol and criteria.

## 9.4 Timing

RAST may be used in any stage in the life cycle of a facility. For new projects, RAST may be used with minimal inputs at an early stage to assist in developing equipment design information, evaluating inherently safer alternatives and assessing the system equipment layout. This is also covered further under Safety Studies (Chapter 3), HAZID (Chapter 5) and LOPA (Chapter 7). For an existing facility, RAST may be used to determine the impact of plant modifications to hazard scenarios to evaluate effects of proposed changes on the risk currently addressed with the existing protection layers.

## 9.5 Advantages, disadvantages and uncertainties

### Advantages

- RAST can be utilised at an early stage of a project before detailed P&IDs are available. However, it does require some knowledge of the final layout and equipment juxtaposition, which may not be available at an early stage.
- RAST assists the study team in identifying risk scenarios by providing a good initial listing that the team may build upon, including adding known incidents applicable to the equipment under review.
- The RAST software may be updated to reflect the specific company process risk requirements and criteria. This is recommended for companies that have an established risk matrix.
- RAST assists the study team in screening for relatively low-risk scenarios where qualitative evaluation of safeguards may be appropriate.
- RAST performs a detailed consequence analysis to identify high-risk scenarios that may be candidates for greater analysis detail such as Layer of Protection Analysis.
- RAST includes Relative Risk Ranking methodology and other techniques to assist the study team in prioritising efforts required to design, implement and sustain the protection layers to effectively manage the risks.

## Disadvantages

- Although minimal input can be used to generate a qualitative risk assessment, detailed input information is required for semi-quantitative risk assessments such as the LOPA.
- Detailed understanding and experience in process hazards and risk analysis methodologies is required to utilise the specialised features of the RAST software, especially when recognising if inappropriate inputs have been used and for interpreting the results. A cross-functional study team is required for effective application of RAST during the review.
- Study results are only considered order-of-magnitude although selected scenarios can be specifically addresses by QRA.
- The listing of hazard scenarios provided in RAST requires the study team to identify additional Initiating Events (or Causes) for some scenarios in addition to identifying scenarios for less common types of operations. However, for an experienced team this is a helpful starting point for 'brain storming'.
- RAST provides a very large list of cause-consequence pairs (scenarios) for consideration in LOPA of which only a few would typically be used in more detailed Risk Analysis. Again, this is not a problem for an experienced team.
- As emphasised in some points above, it is crucial to have experienced process risk personnel to fully understand some of the techniques and methods used within RAST.
- The mathematical simplifications used within RAST must be appreciated, for example that gases are treated as ideal and thermodynamic properties are correlated as simple linear relationship with temperature. This will affect estimates of risks for chemical processes operating near the critical point. Details of all simplifying assumptions are documented in the manuals [1].

## Uncertainties

- The severity of hazard is often difficult to quantify and relies on many parameters with high levels of uncertainty such as lethal dose, reaction rates during upset process conditions and the reliability of human actions or reactions.
- The results are only at the order of magnitude level and any issues, which create concerns, must be subjected to a full QRA. Estimating the number of people impacted for a scenario depends on many uncertain variables, e.g. lethal dose, wind direction, occupancy levels, etc.

# 9.6 Example of a RAST evaluation

## Indoor blending operation for manufacture of inks and dyes

The plant was an 1100 $m^2$ ink and paint manufacturing that began operations in the early 1960s within a minimally populated peninsula. Over several years, a large marina and many family homes were built adjacent to the manufacturing plant with some homes only 50 m away. The facility stored alcohols, heptane, other solvents, pigments, resins and nitrocellulose within and outside of the building.

Solvents, pigments, resins and nitrocellulose were mixed to produce inks and paints in 11 $m^3$ vessels equipped with top-mounted agitators and a steam heating jacket. Mix tanks had a 30 cm diameter access hatch to keep debris from falling into the tank but allowed vapour or air to pass through the opening. The initial mixture of more than 7600 litre of heptane and propyl alcohol was added to the tank from 1900 L totes. Resin was hand loaded from drums to the top of the tank and the mixture was then heated to between 32°C and 49°C to dissolve the resin. Temperature control was achieved by manually opening a 6-mm steam valve leading to the steam heating jacket. Following a quality control check, the liquid was pumped out the bottom of the mix tank to smaller pigment mixers, as needed. Unused resin-solvent mixture would remain in the mix tank until it was all utilised in specific ink products. These data are entered into the RAST spreadsheet followed by the chemical data, in this case for a solvent mixture of equal fractions of heptane and 1-propanol with a small quantity of dissolved solids and operating conditions of 0.01 bar gauge and 40°C.

The equipment information follows: the vessel volume is 11 $m^3$ with maximum working pressure as 0.1 barg. The relief device is essentially the 30-cm access hatch, venting 'Indoors'.

The process conditions are at ambient temperature of 25°C with a maximum flow rate to the tank of approximately 190 L/min from a tote. The maximum liquid height in the vessel is 2.4 m.

**Plant layout**—The plant layout is shown in Fig. 9.5. Only minimal information will be entered into RAST at this time similar to early design stages for a new facility where minimal information is available. In addition to this minimal information it is assumed that during normal work hours, two people might be in the production area and five in the offices and laboratory and ten people located at the marina between customers, maintenance and sales staff.

The initial RAST report gives several suggested scenarios for the formulation mix tank identified through the pre-populated RAST scenario lists as a starting point for the study team. In addition, a 'worst' consequence severity using the specified risk matrix is noted. Table 9.2 shows the highest consequence scenarios provided by the RAST software. RAST

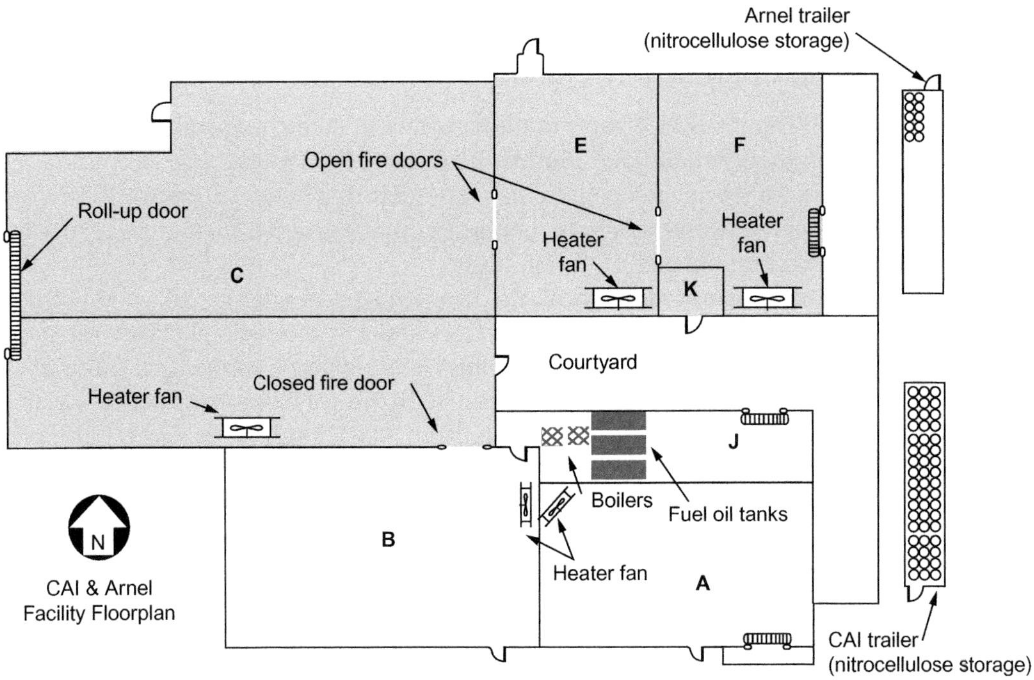


**FIG. 9.5** Building area with production areas highlighted.

recognises that this list may not be complete, providing the study team with a first set of scenario suggestions to build upon. Scenarios that appear to have a potential for internal building explosion and high-consequence severity incidents include excessive heating and overfill.

A consequence summary report may be obtained for any loss event. For the formulation mix tank the *Vapour Vent—Heat Transfer* represents a 'worst' consequence for filling the enclosed area with flammable vapour although the enclosed area concentration is not estimated to reach the lower flammable limit if the ventilation system was running. It is worth noting that the blast overpressure radius estimated in RAST was in close agreement with modelling depicted in the incident report published by the US CSB. Quantifying incident outcome is one of the unique aspects of using RAST.

A Layers of Protection Analysis on this scenario indicates that safeguards are not sufficient to effectively manage this event, as evidenced by the incident. Ensuring the ventilation system was always running, a high-temperature interlock to the steam source, emergency venting of the mix tank to an outdoor location, and procedure to seal the 30-cm solids addition hatch would likely have provided sufficient protective layers to appropriately manage the risk associated with this scenario.

This example is presented in full as one of the case studies provided on the CCPS website [2].

**Table 9.2** Suggested scenarios for consideration by the study team—a selection of a larger listing.

| Scenario type | Scenario comments | Parameters and deviation | Initiating event (cause) | Initiating event description | Loss event | Outcome |
|---|---|---|---|---|---|---|
| Drain or vent valve open | Drain or vent valve left open following infrequent maintenance, purging or cleaning | Flow—loss of containment | Human failure action once per quarter or less | Operator leaves drain or vent open following infrequent maintenance | Drain or vent leak | Flash fire or fireball |
| Excessive heat input—heat transfer | Vapour pressure plus pad gas exceeds maximum allowable working pressure or relief set pressure at ambient or heating media | Pressure—high | BPCS instrument loop failure | Failure of flow control | Vapour relief vent—heat transfer<br>Equipment rupture at saturation temperature | Flash fire or fireball; building explosion<br>Flash fire or fireball; building explosion |
| Excessive heat input—pool fire exposure | Vapour pressure exceeds relief set pressure from pool fire exposure | Pressure—high | IEF = 2 as determined by Process Safety | Leak of flammable material or material above its flash point which may ignite | Vapour relief vent—fire<br>Equipment rupture at fire conditions | Indoor toxic release, flash fire or fireball, building explosion<br>Flash fire or fireball |
| Ignitable headspace | Chemical is flammable or combustible: maximum operating, mechanical energy or heating media temperature exceeds flash point less 5°C | Composition—wrong concentration | BPCS instrument loop failure | Failure of pressure or noncombustible atmosphere control | Equipment rupture—deflagration | Flash fire or fireball |
| Overfill, overflow or backflow | Overfill or backflow of liquid with spill rate equal to the feed rate to a maximum quantity of the available | Level-high or flow-backflow | BPCS instrument loop failure | Failure of level indication with continued addition of material | Overfill release<br>Equipment rupture at operating temperature | Flash fire or fireball<br>Flash fire or fireball; building explosion |

# References

[1] The latest versions are available at: https://www.aiche.org/ccps/resources/tools/risk-analysis-screening-tool-rast-and-chemical-hazard-engineering-fundamentals-chef/rast-user-and-chef-manuals.

[2] Chemical Safety and Hazard Investigation Board Report 2007-03-I-MA, https://www.aiche.org/sites/default/files/docs/book-pages/case_study_-_confined_space_explosion_cai_and_arnel.pdf4, May 2008.

# Further reading

[3] Four detailed examples, including the one illustrated above, are available at: https://www.aiche.org/ccps/resources/tools/risk-analysis-screening-tool-rast-and-chemical-hazard-engineering-fundamentals-chef/case-studies.

# 10 Checklists

Chapter outline



## 10.1 Definition

A checklist is a carefully compiled, comprehensive list of protective measures, procedural steps, material properties, hazards or 'good practice' design features that have been compiled by experienced personnel for a particular application. Checklists are used to systematically check designs, operations, system status, etc. for compliance with known codes, standards or other specific requirements.

Input to this list may come from:

- Global experience
- Task analysis
- HAZOP

## 10.2 Description

Checklists have been generated and used in every phase of process design from conception to decommissioning and is probably one of the first of the Hazard Identification tools.

A Guide to Hazard Identification Methods. https://doi.org/10.1016/B978-0-12-819543-7.00010-0

They are often used as a part of other hazard analysis methods. They can contribute to hazard identification, hazard evaluation, risk analysis and the assessment of protective systems.

A table of references for checklists given by Lees [1] identifies the many areas where lists are available, as well as giving several general references. CCPS [2] includes a long appendix listing questions for many areas of process operations. Further examples are provided by the HSE [3]. It is not possible to give a comprehensive account of checklists here; only the essential characteristics are covered.

A checklist is a list of questions, checks or items on a particular aspect of the task under consideration. A good checklist will have been prepared for a specific application, often a very precise and limited one. Usually there will have been more than one person involved in creating the list and all authors should have extensive experience in that area. Sometimes the response required to each question is a simple Yes/No response, with space for a comment where a negative response is made. Sometimes a precise piece of information is needed, such as the value of a material property, and lists may require some thought before answers can be given.

The main point of a checklist is to assemble questions based upon historical experience of the type of problem or task and to ensure that all have been considered. It is then possible to see whether an appropriate action has been taken or standard has been met.

When it is decided that a checklist analysis is relevant the sequence is to select or generate the list, to perform the review and finally to document and implement the results. In many cases it is possible to take an existing checklist but where this is done it must be examined to ensure it is both relevant and comprehensive. This should be done by staff who have extensive experience in the area under review. If no suitable checklist is available the analyst should generate one based on experience, plant history and references. It is useful to have several people, including senior personnel, involved in the creation of the checklist although less experienced engineers may carry out the analysis. It is advisable to treat a checklist as an open document capable of improvement and extension based on past usage and new experiences.

The analysis using the checklist usually starts with planning how the review will be carried out and how the necessary information can be gathered. The actual analysis may involve plant inspections, checking records and documentation, discussion with operators and observation of working practice and perhaps interviews to gather personal perceptions. For many items the information recorded may be a simple entry to confirm compliance plus some details whenever a negative response is recorded or when there are insufficient data available to complete the response.

A report can then be prepared summarising the deficiencies found together with any suggestions for improvement. It then falls to the responsible person—who should be identified at the outset of the analysis—for decisions on action and implementation. A copy of the completed checklist should be included with the report.

## 10.3 Resource requirements

### Manpower

Checklists are normally completed by one person, occasionally by a small team. It is important that the user is competent in the area being analysed.

### Software

Most lists are done on a hard copy but they can usually be converted onto a spreadsheet or word processing package if required.

### Data

The requirement varies greatly with checklist application. Normally existing plant documentation, procedures and other records will be needed.

### Timing

This depends upon the area of application and the point at which the data will be available. It is usually obvious when the optimum time will be.

## 10.4 Advantages, disadvantages and uncertainties

### Advantages

- Checklists can be customised to the requirements of individual applications and companies.
- Their use is straightforward, well structured, easily understood and can ensure consistency.
- They are very useful for standard or repeated operations to ensure that no basic problem is overlooked.
- Their value is enhanced if analysts are encouraged to add to the list and to maintain an open mind during their use.
- They can eliminate potential problems at an early stage of a project development and prevent them becoming engrained in the detail of the design.

### Disadvantages

- The creation of a good checklist requires considerable expertise and experience. Production and validation can be resource intensive.
- If the checklist is incomplete due to lack of experience a fault may be overlooked.
- The technique can lead to a 'blinkered' study that does not fully explore the hazards associated with the process.

- Use of checklists on their own can lead to a false sense of security.
- The technique is less suitable for new and novel processes.

### Uncertainties

Checklists are not quantitative. Expertise is required to decide if a standard has been met and to make an accurate qualification alongside a negative response. Thus there is some subjectivity in the analysis.

## 10.5 Applications

Practically every aspect of hazard identification and evaluation has been linked to a checklist. Used within their limitations they can be a valuable means of enhancing safety in most operations. Some examples are given:

- Preparation of a code or standard
- The preparation of a work (operating) procedure
- The preparation of a maintenance procedure
- The preparation of a permit to work (this can be quite extensive)
- During certain phases of a capital project where issues can be identified before the project has advanced too far, for example prior to site restitution.

## 10.6 Examples

### Example 10.1

This illustration is based upon an extensive example provided by Scott and Crawley [4] covering process design and operation. They note that with checklists there is a danger of slavish adherence with the user simply seeking exact answers to specific questions. This may show whether the design is adequate in these specifics, but it will not achieve confidence that the overall design is safe. To avoid this the checklist uses generalities in order to stimulate wide-ranging questioning. Only a small proportion of the checklist is reproduced here.

The checklist is subdivided into major areas with many questions listed under each heading. In total there are 196 separate points for the analyst to consider. The headings are:

Conceptual design
Detailed design process
Detailed design layout
Operation and maintenance
  Start-up and shutdown
  Operation
  Maintenance
  Human and instrument error
  Analysis

- Documentation
- Safety
  - Personnel protection
  - Spillage
  - Fire

The questions for the detailed design process are listed below.

*Detailed design process*

1. Determine the quantities and physical states of the process materials in all stages of production, storage and handling. Can the quantities be reduced or temperature and pressure requirements be eased?
2. Are the materials of construction compatible with each other and with the process materials at all temperatures, pressures and concentrations envisaged in the process?
3. Is there a risk that poor mixing or inefficient distribution within reactors or heat sources, either by malfunction or by design error, can give rise to undesirable side reactions, hot spots, reactor runaway, fouling or other problems?
4. Can dangerous materials build-up in the process (e.g. traces of combustibles or noncondensable materials)?
5. Have all aspects of catalyst behaviour been allowed for (e.g. ageing, poisoning, disintegration, activation and regeneration)?
6. Can the process be simplified by the removal of any item? Will this improve safety?
7. In the event of the failure of each item what are the consequences? Is such a failure acceptable?
8. Has the chance of failure been quantified?
9. What would failure of any item mean? Is it desirable to build in redundancy or diversity?
10. Have fault trees been produced for all failure events of interest?
11. Can the reliability, flexibility and availability of the process be improved?
12. Are the effects of a failure in utility supplies minimised?
13. What is the effect of a failure in a related plant that either receives from or supplies chemicals, utilities or both?
14. Which items need installed spares, bypasses or both?
15. Which items require a back up supply of a utility, particularly emergency electrical power?
16. What happens if bypass or isolation arrangements fail? Is positive isolation of fluids required?
17. Has the item or plant been used before in a similar duty? Is it reliable and easy to service?
18. Does the plant item fulfil the requirements of the design specification? If not, could changes to this specification be approved?
19. Does the manufacturer have relevant experience concerning each process unit?
20. What items of equipment are designated as critical plant items?

21. Has the specification been discussed with the manufacturer? Have desirable and undesirable aspects of the tender been agreed?
22. Subject each line on the engineering line diagrams to a hazard and operability study.
23. Are the operating limits modified by safety considerations resulting from operability studies?
24. Quantify the consequences of any serious hazard using failure mode and effect analysis and/or fault trees in conjunction with hazard analysis.

## Example 10.2

A large company that runs a number of batch manufacturing processes, each involving many process steps using multipurpose equipment, carries out a risk analysis for the whole sequence of steps. A customised what-if checklist can be used to assist in this lengthy and detailed search for unidentified hazards. Each step is described in the manufacturing procedure as a defined interaction between men/machine/chemicals/control system; the use of the checklist is to give comprehensive hazard identification for each step. The procedure is first to fill in a tick list identifying the site services used during the step and the aspects of process operations and control associated with it. Each item that has been ticked is then reviewed using questions that highlight known problems and types of deviation. This ensures that the obvious hazards are considered and that the analyst is challenged to think how they could occur. Interestingly some features of HAZOP study and of What-if feature in this checklist.

The site service list will include:

Electricity
Water
Steam
Coolants
Instrument air
Compressed air
Nitrogen
Vacuum
Absorption/adsorption

The operations and control list will include:

Cleaning
Equipment checks
Evacuation/breaking vacuum
Charging
  Solids
  Liquids
  Amounts

- Sequence
- Rate
- Static

Temperature
Pressure
pH
Heating/cooling
Agitation
Reaction
- Exothermic/endothermic
- Catalysts/inhibitors

Refluxing
Separation
- Distillation
- Centrifuge

Transfers
Connections/flexible hoses
Sampling, etc.

In the What-if/Checklist each of these headings is first amplified to ensure its meaning and range is clear. It can be helpful to list under each heading the HAZOP deviations commonly associated it. Then a series of generic What-if questions are posed for every checkpoint to ensure the analyst considers known issues, firstly to decide if they are relevant in this case and, if so, whether they are of importance. A column in the record sheet is ticked for each important situation and a brief description of the worst case is noted. This approach, at its best, not only ensures that obvious hazards are considered but also helps to identify less obvious ones.

The expansion for a service such as 'Nitrogen' reminds the analyst that nitrogen may be used for purging, inerting, degassing, pressurisation, etc. The most likely deviations are no, more of and less of (flow or pressure), elsewhere (leak) and as well as (contamination). Questions include

- Could the nitrogen be supplied at a different pressure?
- Is there potential for the overpressure of the equipment?
- Would loss of nitrogen purge or inerting be immediately hazardous?
- Can oxygen be evolved in the reaction step?
- Could process chemicals back up into the nitrogen supply?
- Could an operator be exposed to an inert atmosphere (open manway, enclosed charging chute or dryer room)?

The expansion for 'Agitation' has the prompt that it refers to process mixing equipment and possible deviations are none, high or low and before or after the sequence. The questions include:

- Can a critical accumulation of reactants occur in this step?
- Is agitation necessary for heat transfer purposes?
- Is the restart after interruption a critical operation?
- Are variations in stirring speed critical?
- Is the material viscosity a function of speed of agitation?
- What happens if the agitator blades are exposed?
- How is agitation monitored and alarmed?

Such a checklist, used by experienced analysts at an appropriate time when the necessary data are available, will show up many of the potential process hazards.

### Example 10.3

Another example, taken from HAZOP: Guide to Best Practice [5], illustrates a checklist for use in Hazard Study 2. This was derived as a result on the experiences of HAZOP studies and is to be found in Chapter 4.

## References

[1] F. Lees, in: S. Mannan (Ed.), Loss Prevention in the Process Industries: Hazard Identification, Assessment and Control, fourth ed., Butterworth-Heinemann, 2012.

[2] CCPS, Guidelines for Hazard Evaluation Procedures, third ed., Wiley, 2008.

[3] Available via the HSE website at: www.hse.gov.uk.

[4] D. Scott, F.K. Crawley, Process Plant Design and Operation, IChemE Rugby, 1992.

[5] F.K. Crawley, B.J. Tyler, HAZOP: Guide to Best Practice, third ed., Elsevier, 2015.

11

# What if?

Chapter outline



## 11.1 Definition

A brainstorming technique in which a group of experienced persons, familiar with the process being studied, ask questions and voice their concerns about possible hazards. The technique is wide ranging and can be unstructured, although the facilitator may produce some guidance based on a preliminary review of the process being studied. More structure can be introduced through the development and use of a checklist of questions and areas of concern. What if? is probably one of the two oldest Hazard Identification Techniques.

## 11.2 Description

The What-If method is well established [1, 2] and widely used, sometimes under other names such as HAZID. It is quite simple and can be very effective, with applications at practically every stage of a process life. The study is carried out by a small, multidisciplinary team consisting of experienced persons. A study will focus on a defined part of the process using layout drawing, P&IDs and other material and process documents and work through the process asking a series of What-If questions. A preliminary list of questions can be prepared before the meeting, either from an established checklist or by the team members, but other questions will be generated during the study meetings. The team should work logically through the process, following the process stream or the batch sequence. The intention is to challenge different parts of the design and to test that all contingencies have been considered. The questioning should be creative with analysis

A Guide to Hazard Identification Methods. https://doi.org/10.1016/B978-0-12-819543-7.00011-2

of consequences, ranking of criticality and recommendations being agreed by the team members. While, on the face of it, 'What if?' might appear to be a random process in reality it has much of the HAZOP technique within it. 'What if there is ***no flow***?' ***No*** is the guide-word in HAZOP and ***flow*** is the parameter. The Facilitator can add shape to the discussion by asking secondary questions and so steering the lateral thought process. This is a typical tool used in Hazard Identification.

A study of a transfer line could consider questions such as:

- What if there is no flow in this line? What could cause it and what will be the effect?
- What if access if required for maintenance? Can the equipment be accessed?
- What if there is a fire? Can the emergency services gain access? What local facilities are available?
- What if an operator is injured? Will rescue be possible?
- What if an abnormal condition arises?

In each case the team will assess the consequences and, considering the likely causes, give a criticality rating to the event, for example by using a simple consequence-frequency matrix. When it is appropriate, the team should identify the available options or make recommendations for change. For each What-If question the analysis is recorded in a simple table with columns for the What-If question, the Consequences, the Safeguards, the Criticality and the Recommendations.

If the technique is applied early in the process development, for example at the conceptual stage, the questioning will be higher level and less process-risk oriented. Questions would be of the type:

- Environmental security—what if there is a major spillage; what is the effluent and how should it be disposed; what if there is no suitable disposal site?
- Reactor kinetics—what if there is a runaway reaction?
- Personnel—what if trained staff cannot be obtained?
- Neighbourhood reaction—what might cause objections to the development?

A recently developed format is a Structured *What-if Technique* (SWIFT). In this the leader has very specific issues for review plus others that evolve as the study progresses. There are some similarities to the What-If checklist study, but the SWIFT approach is more focused on the specific issues. The technique can be used to follow the layout diagram where issues such as access can be discussed or a P&ID using a modified HAZOP approach. In the former case the leader will discuss the main safety and operability issues with a small team of experienced persons and draw out a structured listing of issues such as:

- Access: Maintenance; Emergency Services; Process Valves; Means of Escape/ Entrapment.
- Loss of Containment: Corrosion; Erosion; Impact; Fire; Blast; Over pressure.
- Escalation Due to: Fire; Blast; Loss of Containment.
- Impact on: Personnel; Escape; Emergency Services.

The discussion can be developed along one line or inter woven with two or more issues such as Loss of Containment, the Impact on Personnel and Means of Escape.

In the second case, based around a P&ID, a full HAZOP study would remain the preferred approach but SWIFT can be used as a pre-HAZOP study filter during the development of the P&ID. In this a high-level approach is taken and issues such as the following can be discussed:

- Integrity of isolation.
- Quality of instrumentation.
- Relief systems—Do they match the likely credible demand?
- Do piping specifications reflect the likely maximum system pressure at upset conditions?
- Are high- to low-pressure interfaces handled adequately?
- Are start-up and shutdown facilities supplied?

## 11.3 Resource requirements

*Manpower*: A What-If study requires between 4 and 6 members in the team. A leader/facilitator and a recorder are required for a major study but the leader/facilitator can also act as a recorder in a small study. The make-up of the other members will depend upon the study and the timing in the project. A team membership of about 6 is optimum for a larger project. It is important to have a mixture of disciplines and for the members to have extensive design, process, operational and maintenance experience.

*Data*: No quantitative data are required. It is essential to have relevant P&IDs, PFDs, operating procedures, etc. available. Also a review of an Accident Database can assist in the development of the structure of the study.

*Software*: Specific software programs are available and other standard programs can be easily adapted for What-If recording. However, discussions can easily be recorded by hand. However, it may be beneficial for QA purposes to have a computerised actions tracking system.

## 11.4 Timing

A study can be done at any time but, obviously, should be early enough to allow the recommendations to be implemented cheaply and easily.

## 11.5 Advantages, disadvantages and uncertainties

### Advantages

- The technique is very flexible and can be used at any part of a process life cycle by changing the list of issues to be discussed.

- The technique is simple and a study can be conducted relatively quickly.
- The technique allows the use of the imagination and knowledge of the group.
- Compared to a HAZOP study, a What-If study can use a different grouping of skills and explore risks that are less process oriented.
- What-If can be very useful in the early stages of a project to identify major issues and areas needing further study.
- A correctly structured study can identify novel problems.

### Disadvantages

- The quality of the technique is very dependent on the skills of the leader, the derivation of the issues for discussion and the knowledge base, the imagination and the intuition of the team.
- If the study is loosely structured the team may waste time on trivial events or miss important areas.
- The results are qualitative and less detailed than other techniques.
- The leader has to carry out some initial research to identify the issues and prompts to start the discussion and to maximise its coverage.
- The quality of the study is difficult to assess.
- Used alone, a study can give a false sense of security.

### Uncertainties

- Not all problems will be identified.
- Without direction and structure the identification may revert to a form of Audit and lack detailed investigation.

## 11.6 Applications

The technique can be used for assessing the issues in a conceptual design through to the issues associated with the construction of a plant and the criticality of deliveries. Indeed, with appropriate change of focus, it can be used at any time in a process life cycle. A study can be very effective when it is not just process directed.

## References

[1] Guidelines for Hazard Evaluation Procedures, second ed., CCPS, 1992. ISBN 0-8169-0491-X.

[2] S. Mannan (Ed.), Lees Loss Prevention in the Process Industries, fourth ed., Butterworth-Heinemann, 2012.

# 12 Failure modes and effects analysis (FMEA) and failure modes, effects and criticality analysis (FMECA)

Chapter outline



## 12.1 Definition

A FMEA considers in a systematic and rigorous manner all of the possible single failure modes of the individual plant items and elements in a system. The effects of each failure on the item itself and on the rest of the system are identified and recorded. The analysis is usually carried out by a small team who should have detailed knowledge of the equipment. In a FMECA the criticality of the consequences and likely frequency are also assessed and recorded.

## 12.2 Description

FMEA is a thorough but time-consuming method of hazard identification. It takes a selected part of a system, usually a piece of hardware, and aims to examine every failure mode of every item and element within it. For each failure mode the consequences must be determined so that the adequacy of the response to this failure can be decided. It is a well-established method and there is ample guidance available in the literature [1–3]. The pattern of analysis is first to define the boundary for the study, decide the depth of analysis

A Guide to Hazard Identification Methods. https://doi.org/10.1016/B978-0-12-819543-7.00012-4

and then to obtain necessary documentation. There must be a defined physical boundary that is usually the physical boundary of the hardware item. Any interfaces with the rest of the system, for example the services, must be identified. A full description of the system, its operation and its situation, similar to that needed for a HAZOP study, is required. Key items of documentation will include the equipment drawings and manuals and information on protective systems so that effects of failures can be fully evaluated. In a full examination it will also be necessary to look at the possible failures of the services (electrical, instrument air, steam, cooling water, etc.). To cover these adequately it is essential that the team members have knowledge of the system, including its operational history and failure modes.

Next a small team, typically the leader and two others with a suitable range of technical knowledge and experience, works through the system in a logical and systematic manner, for example working downstream item by item. A good team would include an engineer familiar with the design and operation of the system and an instrument and control specialist. The leader would take responsibility for the organisation, recording and reporting of the study. The leader should also contribute to the technical analysis.

For each item the team should consider its function and identify every possible failure mode. They then consider the consequences, both for the item itself as well as for the system as a whole. The analysis and recording can vary in detail; the usual headings include:

| | |
|---|---|
| • Component | Identifier and description; function. |
| • Failure mode | As well as fail safe, fail danger, fail degraded and fail neutral, the categories may include failure to start, failure during operation, failure to stop and premature operation. |
| • Effects | This should cover the item itself and other components, as well as the effects on the whole system. |
| • Safeguards | Either here, or separately, the means of detection may be recorded. Compensating provisions against failure can also be noted. |
| • Actions | The comments and recommendations of the team. |

Some of the above headings may be treated separately.
Additional headings can be:

| | |
|---|---|
| • Failure nature | Revealed, unrevealed, degraded or incipient. |
| • Cause | It is not essential to record this but it can be helpful and it will certainly be needed if failure frequencies are to be estimated. Categories normally include specification, design, manufacture, installation, operation, maintenance, environment and outside forces. |
| • Frequency | The estimated frequency of failure must be estimated if a criticality analysis is being done (FMECA). A 4 or 5 point scale is normally used. |
| • Hazard category | A ranking of the consequences on a 4- or 5-point scale. This may be done separately for different types of consequence, e.g. injury/death, production loss, equipment damage, environment, etc. |

In a FMECA the allocation of the frequency and hazard category values is most efficiently done at the end of a session to encourage consistency. Using a $4 \times 4$ or $5 \times 5$ grid, a criticality rating can be assigned to each failure mode.

## 12.3 Resource requirements

*Manpower*: An experienced analyst or, more beneficially, a small team of engineers, one of whom may act as scribe, with a thorough understanding of the design and function of the system and its components.

*Data*: A detailed set of diagrams with supporting documentation to provide a full and clear description of the system and its workings. For a FMECA, failure rate and consequence data are needed. As with the FTA, obtaining high-quality failure rate data may be difficult and, in order to work with valid data, it may be necessary to work at a general level rather than at a very detailed level.

*Software*: A number of dedicated software aids are available to help with the recording of the study but a simple spreadsheet can also be used.

## 12.4 Timing

FMEA may be used for detailed analysis during the equipment selection or the design stages of a new project. It may also be used to assist in the understanding of an accident.

## 12.5 Advantages, disadvantages and uncertainties

### Advantages

- It is an effective method for the identification of hazards associated with mechanical and electrical equipment failures, including reliability problems.
- It is not difficult to apply and the results are easily understood.
- The analysis can highlight both local and general system failures.
- An FMEA on an engineering item is the equivalent of a HAZOP study of a P&ID.
- The manpower requirements are less than those for a HAZOP study.
- A semi-quantitative ranking of the hazards can be produced.

### Disadvantages

- The technique is not the most effective at identifying combinations of failures or when the hazard comes from the whole process although a good team can be expected to identify some combinations.
- The technique concentrates on the equipment and does not address operational errors.

- The analyst(s) must be familiar with equipment functions and failure modes and be able to derive the effects on other sections of the system.
- When applied by a single analyst the technique may miss important interactions.

### Uncertainties

- The value of a FMECA will be limited if the available data are of poor quality and lack robustness or relevance.

## 12.6 Applications

The potential range of applications is very wide, ranging from a small item of equipment to a process unit containing many components. It can be very useful in investigation of incidents or in the analysis of a new design. However, as it is a time-consuming method it should be used with discrimination.

### Example 12.1

The following is an example of a FME(C)A in operation. The numbers used have been chosen as realistic but are not actual values from databases. Deliberately, a very simple example has been chosen. This results in a short, easily understood table, but one that still illustrates the insights this method can give.

The example is based on the most simple and easy-to-understand piece of engineering; a nut and bolt used to close a flange in a pipe carrying propane. The nut has two elements, the body and the thread. The bolt has three elements, the head, the shank and the thread. In general a FMECA analysis would take the whole nut plus bolt as a unit; however, for illustrative purposes it has been subdivided into five elements.

In Table 12.1, columns A, B, C, D and I are normally to be filled in but columns E, G and H will be needed for the Criticality analysis and column F may be used for assessment of improvements.

This analysis shows that the nut and bolt as a unit are not 'critical' but that they should be treated with the appropriate skills and care. Failure is unlikely to be catastrophic provided that the pressure testing is carried out properly.

Another Example is based on a real event over 30 years ago and so does not refer to Buncefield.

### Example 12.2

The level measurement in a liquid storage tank operating at atmospheric pressure utilised a follower or float located in a vertical guide tube perforated every 0.5 m. The location of the follower or float was detected acoustically and a signal sent to the control centre. (Variants will include a measurement by storage of tape or wire on a reel.) The detector was sealed at atmospheric pressure.

| A<br>Component | B<br>Failure Mode | C<br>Effect | D<br>Safeguards | E<br>Failure Nature | F<br>Cause | G<br>Frequency | H<br>Criticality Category | I<br>Action |
|---|---|---|---|---|---|---|---|---|
| 1 Nut—head | Flats damaged during tightening.<br>Fails danger or neutral. | 1 Nut may not be fully tight. Flange may leak process fluids, so resulting in a small fire.<br>2 Nut and bolt cannot be untightened | 1 Pressure test after tightening.<br>2 Nut can be cut off.<br>3 ESD | 1 Revealed<br>2 Degraded. | Poor fitting and supervision. | $10^{-5}$ per bolt. $10^{-1}$ per plant year. | Medium<br>2 | 1.1 Ensure that the correct tools are used during maintenance.<br>1.2 Ensure that the pipe fitters are properly trained.<br>1.3 Ensure that all joint broken are pressure tested to the full operating pressure. |
| 2 Nut—threads | Corrosion due to crevice corrosion.<br>Fails danger. | Joint starts to leak initially as a slow drip but then worsens. | | 1 Revealed | Rain water attack. Condensation on cold surfaces. | $10^{-2}$ per plant year. | Medium<br>2 | 2.1 Use "thread guard" or grease to protect the threads. |
| 3 Nut—threads | Overtightened<br>Fails danger | Joint integrity degraded.Joint may leak on line. | 1 Pressure test after fitting.<br>2 ESD | 1 Degraded<br>2 Danger | Poor torque control. | $10^{-2}$ for large bolts per plant year. | Low<br>1 | 3.1 Use preset torque "impact wrenches." |
| 4 Nut—threads | Cross-threaded.<br>Fails danger. | Joint cannot be tightened. | 1 Pressure test after fitting.<br>2 Nut can be cut off.<br>3 ESD | 1 Revealed<br>2 Degraded | Poor fitting. | $10^{-1}$ per plant year. | Medium/low as the fault will be obvious to the pipe fitter.<br>1 | 4.1 There is no real action other than to use experienced craftsmen. |
| 5 Nut—threads | Galled.<br>Fails safe. | Nut cannot be removed when required. | ! Nut can be cut off. | Unrevealed initially but revealed at maintenance. | Poor materials specs. | $10^{-1}$ per plant year | Low<br>1 | 5.1 Consider the benefits of "thread guard" and review the piping material specification for detail on galling. |
| 6 Bolt—shank See 3 above | Over tightened and the bolt is yielded. | See 3 above | See 3 above | See 3 above | See 3 above | See 3 above | See 3 above | 6.1 See 3 above |
| 7 Bolt—head | See 1 above | See 1 above | See 1 above | See 1 above | See 1 above | See 1 above | See 1 above | See 1 above |
| 8 Bolt—thread | See 2 above | See 2 above | See 2 above | See 2 above | See 2 above | See 2 above | See 2 above | See 2 above |
| 9 Bolt—thread | See 3 above | See 3 above | See 3 above | See 3 above | See 3 above | See 3 above | See 3 above | See 3 above |
| 10 Bolt—thread | See 4 above | See 4 above | See 4 above | See 4 above | See 4 above | See 4 above | See 4 above | See 4 above |
| 11 bolt—thread | See 5 above | See 5 above | See 5 above | See 5 above | See 5 above | See 5 above | See 5 above | See 5 above |

*fitting*, workmanship; *specs*, specification; *ESD*, emergency shutdown.

**Table 12.2** FMEA for a float level measurement.

| A<br>Component | B<br>Failure Mode | C<br>Effect | D<br>Safeguards | E<br>Failure Nature | F<br>Cause | G<br>Frequency | H<br>Criticality Category | I<br>Action |
|---|---|---|---|---|---|---|---|---|
| 1 Follower | Sticks in guide tube<br>Unrevealed | Level not detected<br>Tank overflow or low level | Non | Danger | 1 Rust in guide tube<br>2 Too tight fit of follower in tube | Unknown | High | 1.1 Devise a credible test procedure<br>1.2 Fit a diverse level measurement<br>1.3 Check the installation as fitted for clearances<br>1.4 Check for weld protrusions and burrs in tube<br>1.5 Ideally use drawn piping |
| 2 Follower | Is holed and sinks<br>Unrevealed | Indicates low level when the tank may be overflowing | Non see actions above | Danger | Holed by corrosion or wear | Unknown | High | 2.1 As above<br>2.2 Review materials of construction |
| 3 Follower | Reaches the top of the perforations in the guide tube<br>Degraded | Level does not appear to change as the gas is compressed into a dead pocket<br>Tank overflows | Non | Danger | Incomplete perforating of guide tube | Unknown | High | 3.1 Inspect the guide tube when fitted in place<br>Treat the perforations as a high priority<br>This might also be identified during a HAZOP |
| 4 Guide tube | Perforation incomplete<br>Degraded | A s above | As above | As above | As above | As above | As above | As above |
| 5 Guide tube | Corrosion<br>Degraded | Level not detected<br>Process upset | Non | Danger | Follower jams in corrosion products | Unknown | High | See 2.2 |
| 6 Acoustic detector | Wear and tear<br>Degraded, unrevealed | Level not detected<br>Tank overflows or process upset | Non | Danger | Level goes 'blind' | Unknown | High | See 1.1 and 1.2 |
| 7 Acoustic detector | Damaged due to impact<br>Degraded, unrevealed | Tank overflows or process upset | Non | Danger | As above | Unknown | High | 7.1 review physical protection to detector. Should it be fitted into a protective box |
| 8 Acoustic detector | Damaged power supply | Tank overflows or process upset | Non | Danger | Level goes 'blind' | Unknown | High | 8.1 review physical protection to power supply and also reliability of supply |
| 9 Signal transmitter | As above | As above | Non | Danger | As above | Unknown | High | 9.1 See 7.1 and apply to transmitter<br>9.2 See 8.1 and apply to transmitter |

Graphically the level measurement would look as shown. There are only four components, the guide tube, the follower or float inside the guide tube and the acoustic detector and the transmitter to the control centre (Table 12.2).

As indicated above this was a real event and the guide tube perforation was incomplete. The top 1 m of tube was not perforated resulting in a tank over flow. While with hind sight the FMEA has produced the correct answer it would have generated questions if applied before the event.

Actions 1.1, 1.2, 1.3, 1.4, 1q.5, 7.1, 8.1, 9.1 and 9.2 are all good practice.

## References

[1] S. Mannan (Ed.), Lees Loss Prevention in the Process Industries, fourth ed., Butterworth-Heinemann, 2012.

[2] G. Wells, Hazard Identification and Risk Assessment, IChemE, Rugby. ISBN: 0-85295-353-4, 1996.

[3] BS 5760 Part 5, Guide to Failure Modes and Criticality Analysis, ISBN: 0-580-19660-7, 1991.

# 13

# Fault tree analysis

Chapter outline



## 13.1 Definition

Fault tree analysis (FTA) is a technique that focuses on a particular undesired event or main system failure (the top event) and aims to determine all of the ways in which it could occur. The fault tree displays graphically the various combinations of equipment failures and human errors that may lead to the top event. It can be used to identify the root causes of a hazard and, with suitable data, can also be used to assess the probability/frequency of the top event. Fault tree analysis is often used with consequence modelling in Quantified Risk Assessment (see Chapter 15).

## 13.2 Description

Fault tree analysis is a specialised method that is used selectively in hazard identification. It can be very useful in identifying the root causes of a major hazard that has already been identified and which would only occur under quite complex conditions. Traditionally the fault tree is constructed vertically with the final outcome at the top—'the top event'. The

A Guide to Hazard Identification Methods. https://doi.org/10.1016/B978-0-12-819543-7.00013-6

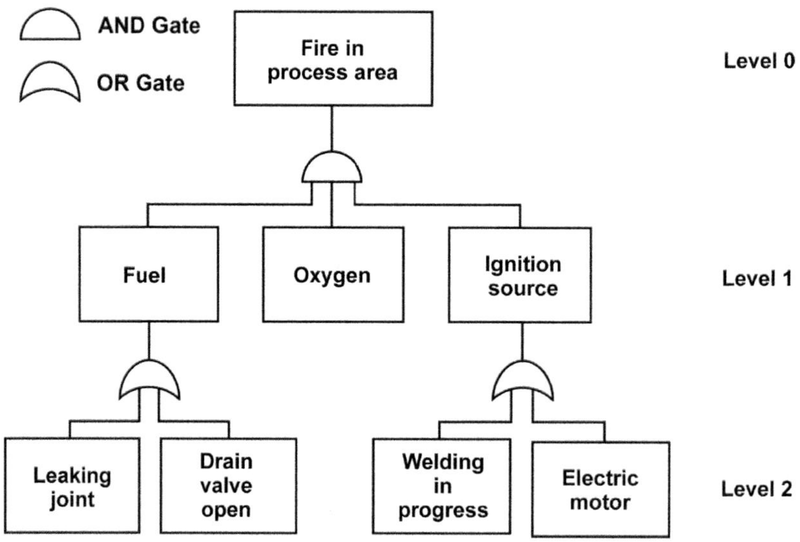


**FIG. 13.1** Simple fault tree for a process plant fire.

fault tree gives a visible structure of these conditions and, by quantification using frequencies and probabilities, an estimated frequency for the top event. The tree can be constructed 'top down' building up the causes to the top event or 'bottom up' working from causes to the top event. It usually finds use in the 'top down' approach.

In a completed fault tree the top event is linked to the initiating events through a series of intermediate levels where the necessary conditions for the propagation towards the top event are combined in AND or OR gates [1, 2]. In a large tree there will be many levels, which grow in complexity and detail as the analysis proceeds away from the top event.

A simple fault tree for the occurrence of a fire in a process area would start with definition of the top event, determination of the essential conditions and then consideration of how these might arise. Such a fault tree is shown in Fig. 13.1.

In this tree the top event is put at level 0. At level 1 the essential combination required for a fire is set out and linked through an AND gate, since all three conditions must exist to initiate a fire. The requirement for oxygen needs no further development as air is always present external to process equipment. The other two conditions, excluding oxygen, at level 1 could arise in a number of ways. Two each are given at level 2 for fuel and for ignition source. They link through an OR gate since they are alternatives and only one need exist to give the level 1 condition. It would be easy to suggest further entries for level 2.

It is evident from this simple example that the construction of a fault tree is not a trivial task. In fact a systematic approach is essential, for example by starting with a generic tree as in Fig. 13.2.

The construction of the tree should start with a full and precise definition of the top event. Wherever possible, development should proceed down a level at a time, with each level being fully developed before the next one is considered. There should be clear

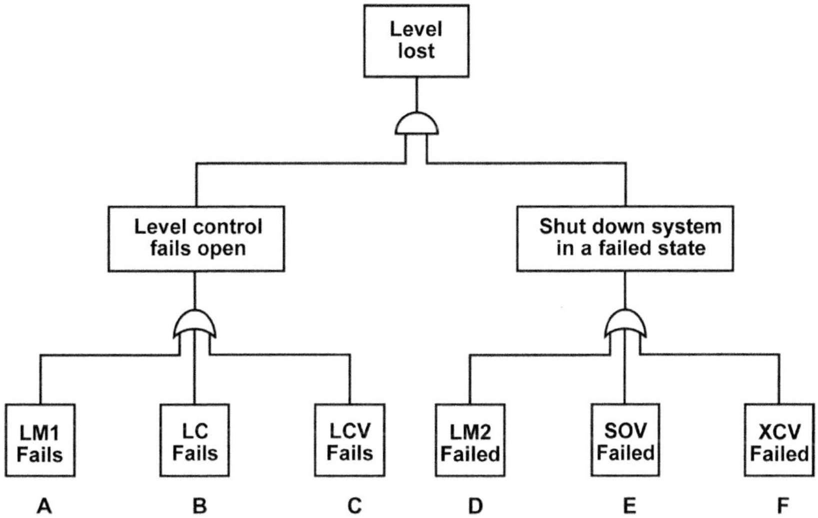


**FIG. 13.2** Generic fault tree for a 'top event'.

statements at each level in an event box of the fault condition at that point. Gate inputs should be linked to defined fault events, not into other gates. Proceeding slowly and thinking clearly are helpful characteristics in an analyst, in addition to experience in the method and knowledge of failure modes and frequencies. Any conclusion from FTA may be seriously in error if either the fault tree is incomplete or its logic and layout are incorrect.

Before quantification the fault tree should be analysed to eliminate double counting, for example deriving the 'minimum cut set' by Boolean algebra. Thus for an event such as a vehicle hitting a pipeline it may simultaneously give the conditions for presence of fuel and an ignition source. Only one entry is needed in the fault tree for this particular event. The example at the end of this section gives a simple illustration of minimal cut sets. More detail and alternative methods are given elsewhere [1, 2].

A fault tree can be quantified by assigning a frequency or a probability to every primary item, starting at the bottom of the tree. These can be combined to give an estimate for the frequency of the top event. The simple rules that apply to AND OR gates are set out in Table 13.1.

Where probabilities combine at an OR gate the output can often be adequately approximated to $P_A = P_B + P_C$ as the product $P_B P_C$ will be small compared to the other terms if each probability is 0.1 or smaller. If either of the nonallowed combinations is present in a fault tree then its layout is incorrect. The rules for combination can be presented using Venn diagrams, as shown by Wells [3]. This is helpful when more than two inputs exist.

An essential factor in the development and quantification of a fault tree is the correct inclusion of common mode failure (CMF; Section 13.6). The failure rate data for any instrument or item of equipment include both 'wear and tear' and CMF contributions. CMF effects can arise from errors in design, manufacture, installation, maintenance or testing and may result in the simultaneous or near simultaneous failure of two identical

**Table 13.1** Valid combinations of frequency (F) and probability (P).

| Gate | Symbol | Meaning | Boolean algebra relation | Inputs | Outputs |
|---|---|---|---|---|---|
| AND | 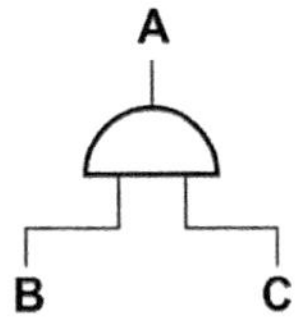 | Output exists only if all inputs exist | $A = BC$ | $P_B$ AND $P_C$<br>$F_B$ AND $P_C$<br>$F_B$ AND $F_C$ | $P_A = P_B P_C$<br>$F_A = F_B P_C$<br>Not allowed |
| OR | 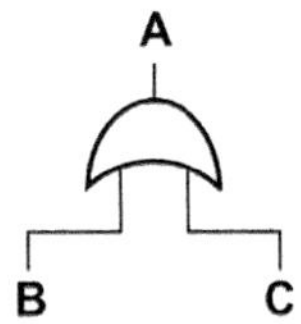 | Output exists if one of more inputs exist | $A = B + C$ | $P_B$ AND $P_C$<br>$F_B$ AND $P_C$<br>$F_B$ AND $F_C$ | $P_A = P_B + P_C - P_B P_C$<br>Not allowed<br>$F_A = F_B + F_C$ |

elements. This can have extremely serious consequences when spare or redundant items are included in a design, in diverse shutdown systems and in majority voting systems (High-Integrity Protections Systems, HIPS). In the absence of direct data a reasonable assumption is that CMF represents 10% of the total failure rate. If the construction and quantification of the fault tree is done on the basis that all the failures are simple random events then there is likely to be a significant overestimate of the reliability of the system and hence an underestimate of the frequency of the top event. CMF must be drawn out as a discrete element of the FTA as shown in the example. There are strategies for overcoming CMF but the effects can never be totally eliminated.

The final check is to ensure that the result of the fault tree has units probability or of frequency 'Per Year' and not $(\text{Per Year})^2$. The analysis of each gate should produce only a frequency or probability. Both can be additive an 'OR' gate but only probability and frequency can be combined in an 'AND' gate.

## 13.3 Resource requirements

*Manpower*: The development of a fault tree is usually carried out by one person but it can be useful to have assistance in defining the entries to the gates. The developer must be skilled in the use of fault trees, understand the likely contributors to the top event and be able to analyse the data in a critical manner. It is sensible to have an independent critical review of the final version of the fault tree.

### Software

Writing out a fault tree and its reduction to the minimum cut set can be slow if done manually and so a software aid can be beneficial. Several are listed in reference by CCPS [2].

### Failure rate data

The data used in a fault tree must be robust and relevant. Robustness comes when a datum is derived from a large number of items and failures, which reduces the statistical uncertainty. Relevance means that the sample is similar to the study under review, taking into account maintenance standards, operating severity and working environment. Items taken from a Data Bank must be critically reviewed to ensure they meet these standards [see Risk Assessment (Chapter 15)].

### Probabilities

Values are normally based on experience and historic data. Some, such as human error, may be industry specific. Also probabilities may be a function of the size of the original event, such as the toxic or flammable release rate and condition. The probability of failure of a trip or other protective system is usually stated as a fractional dead time (FDT). In most cases the major part of this is calculated as $\frac{1}{2} \cdot FT$, where $F$ is the frequency of fail to danger faults and $T$ is the test interval. The full calculation of FDTs is covered by Lees [2], also including the contribution of human factors.

## 13.4 Timing

A fault tree analysis is usually carried out early in the project development. If the process integrity is based on a major piece of shutdown logic the most likely time is after the concept selection Study 1 of Process Hazard Studies (Chapter 4) but before the detailed design. If the instruments are more standard or simplex it might be appropriate to delay the analysis until the detailed design is complete and has been reviewed by HAZOP study. Timing may be critical as should there be serious issues, which are revealed at a later time there could be serious cost implications particularly if High-Integrity Protective Systems are required as a corrective action when it might have been easier to design the issue out a Study 2.

Occasionally the need for analysis may arise from the HAZOP study itself or be used to support the concept selection where critical issues have to be dealt with.

## 13.5 Advantages, disadvantages and uncertainties

### Advantages

- Fault tree analysis is a deductive technique focussed on the identification of combinations of failures, equipment and human that can lead to a hazard.
- It is a structured and methodical process that leads to a clear and logical graphical description of the root causes of a hazard.

- The fault tree can be used to identify key or single point dependencies leading to the hazard using the minimum cut set technique.
- The fault tree assists in the identification of the most critical elements leading to a hazard, particularly if the result is numeric. Numeric answers can be 'sensitivity tested'.
- Use of the technique may encourage the analyst to explore novel or new causes that might contribute to an unforeseen hazard.

## Disadvantages

- The construction of fault trees is time consuming and should only be used in the most appropriate circumstances. Each fault tree should refer to a specific problem that has often been identified by another hazard analysis method.
- The analyst needs to be skilled in the use of fault trees, particularly in determining the degree of detail to be included, and have a detailed understanding of the system under analysis.
- The value of a study will be limited if the available data are of poor quality and lack robustness or relevance. It is not a means of getting a precise prediction from imprecise data. Poor data may lead to an incorrect ranking of the risks.
- All results must be treated with care.
- The results should, for major events, be conditioned by consequence analysis and risk criteria.

## Uncertainties

- All of the data used has a spread of values, often expressed as a standard deviation. These can be combined at the AND and OR gates using the usual statistical rules to estimate the uncertainty in the final value.
- It is essential that full allowance is made for common mode failures and it may be necessary to make an arbitrary allowance for the possibility of these.
- There are likely to quite large uncertainties in values assigned to human factors, perhaps 20%–100%, since judgement is always needed and the values may be training and site dependant.
- The estimates for the top event are likely to be meaningful to 1, or at the most 2, significant figures. FTA is not a means of obtaining a precise result from imprecise data.

## Applications

- The most appropriate use of fault trees is in the analysis of highly instrumented systems, for example high-integrity trip systems and shutdown systems. In particular the analysis is good at identifying single cause dependencies and common mode failures.
- The technique encourages exploration of remote but credible causes of a hazard.
- The method is valuable when a numerical answer is needed.

## 13.6 Failure rate or reliability data and common mode (cause) failure

After due consideration it was felt that this section fitted better into this chapter as opposed to Risk Assessment (Chapter 15) more particularly as there is a figure (Fig. 13.6) which refers to these modes.

There are a number of sources of equipment failure rate data and human performance (failure) data particularly under stress and confusing information. It is for the user to select the correct source and with great care as it is likely that the data will be case specific so no hard and fast rules can be given. As an example of this a pump seal on water duty may leak with a tolerable drip and still be considered to be 'available'. However, the same leak of corrosive or toxic material would require an immediate shut down and the pump declared 'nonavailable'. The data bank may not differentiate between the two fluids. The user has to make the judgement as to the suitability of the data. Likewise internal corrosion will be case specific but external corrosion (corrosion under insulation, CUI) may be due to the local environment or working practices.

Leak data are to be found in databases usually in the form of size and frequency for any specific item (pump, vessel, heat exchanger and per unit length of piping.) This must be viewed critically to ensure that it is case specific.

It is important to recognise early on in this section that failures of equipment and humans do have some common factors. This means that duplication is not truly a case of a spare as that item may also have a common or cause failure mode. The two modes are deliberately separated out but usually are referred to as 'mode'.

Two items manufactures by the same company may have a common mode failure. These could be due to:

1. A poorly manufactured or fitted component
2. Poor design
3. Poor installation on a site
4. Poor maintenance on site
5. Lack of maintenance on a site
6. Abuse

Common cause failure could be due to a harsh operating environment:

1. Process fluids (erosion or corrosion)
2. Heat
3. Vibration
4. Impact
5. Humans and abuse

There is a second common cause, which will be addressed under Audits (Chapter 17), namely poorly supervised maintenance and procedures and poor training. These are Safety Management Systems, which the assessor should bear in mind.

There is a general rule of thumb that common mode/cause failures represent a fixed percentage of all failures. This is given a symbol $\beta$. It has been suggested that $\beta$ should have a value of 10%. The use of this value would indicate that redundant, majority voting systems such as High-Integrity Protective Systems would be too unreliable. In reality it will be nearer 2% or 3%.

Likewise there are tables for human failure rates under different stress regimes. These are not detailed but are available as the failure rate may be related to training and other features in a design. Values can be as high as 0.5 and as low as 0.001. These may affect the final, outcome of the study.

**Never underestimate the impact of common mode/cause in any risk assessment. It is potentially the most important part of the assessment particularly with majority voting systems.**

## Shut down systems

Shut down systems (Emergency Shut Down—ESD) are a key part of Risk Assessment (Chapter 15) and fits better in this chapter. It is imperative that these systems are maintained properly, tested properly and the performance is compared to that used in the risk assessment itself (an original assumption). If the test results fall short of the targets corrective actions are required.

There are three elements in the reliability of the shut down system:

- The probability that the system itself is in a failed state.
- The probability that the system was not returned to an operational state following function testing.
- The probability that the system was under test when there was a demand upon it.

Modern processors have internal checking systems; however, there are two key components, which cannot be fully checked 'on line', the detector and the final operator. If the system is not tested fully the likelihood that the components in that system have failed (aged) increases with time but against that the more testing the greater is the likelihood that the system will be under test when there is a demand. For simplex systems the optimal test interval is about 2 or 3 months. In specific cases a duplicated set of actuators and valves, with function test overrides, may be justified. This is a case-specific study.

The combination of inability to perform as required is has a number of titles:

- Fractional Dead Time (FDT)
- Probability of Failure on Demand—PFD as in LOPA (Chapter 7)

The probability of the hardware of a protective system being in operation at any time ***T*** years assuming random failure—i.e. no '**burn in** or **wear in**' or '**burn out** or **wear out**' is:

$$e^{-FT}$$

Where:

***F*** = the **sum** of the failure rates of **ALL** of the elements (per year). This is usually obtained from **Failure Databases**. However many databases give the value of *F* as the **total failure rate**. In reality some of the failures are '**fail *safe***' or '**spurious**', that means that the shutdown system fails in a safe manner and shuts the process down. This is often given a failure rate designated as '**S**'. The '**fail danger**' is the other failure mode which is the one of interest where the failure results in the nonoperation of the system on demand.

*T* = the test interval—value in years (every 6 months = 0.5 years).

*Note*: *T* will usually be less than 1.

Therefore the probability of the trip being in a failed state or nonfunctional after *T* years, where *FT* is less than 0.1 is:

$$1 - e^{-FT}$$

The expansion of the exponential equation—$1 - e^{-FT}$ reduces to:

$$FT$$

The average state where *T* is small is:

$$\tfrac{1}{2}FT$$

## 13.7 Example

Some of the main points covered above are illustrated in this simple example. It is of a simple shutdown system that prevents the total emptying of a process vessel. The relevant sections of the P&ID are shown in Fig. 13.3.

The level control system consists of level measurement (LM1), a level controller (LC) and a level control valve (LCV). The low-level shutdown system consists of independent-level measurement (LM2) and a solenoid valve (SOV), which closes the shutdown valve (XCV). Level is lost when excess outflow is not prevented by the shutdown system. The fault tree for this is shown in Fig. 13.4.

The basic events have been labelled A–F. A–C will be frequencies whilst D–F will be probabilities. At the AND gate leading to 'Level control valve fails open' the combined frequency is $F_A + F_B + F_C$. For the AND gate leading to 'Shutdown in a failed state' the combined probability is, to a good approximation, $P_A + P_B + P_C$. The OR gate leading to 'Level lost' requires the multiplication of these two groups, effectively $(A + B + C) \times (D + E + F)$. Thus the combinations are: AD, AE, AF, BD, BE, BF, CD, CE and CF.

To illustrate quantification it will be assumed that all the frequencies are 0.1 year$^{-1}$ and that the probabilities, a combination of failure rates and the test interval, are 0.005. Using these numbers gives a frequency for the top event as 0.0045 year$^{-1}$ or, on average, once in 222 years. However, this is almost certainly in serious error due to lack of consideration of common mode failure (CMF) as will be shown below. Additionally the basic principles of reduction to the minimum cut set will be illustrated.

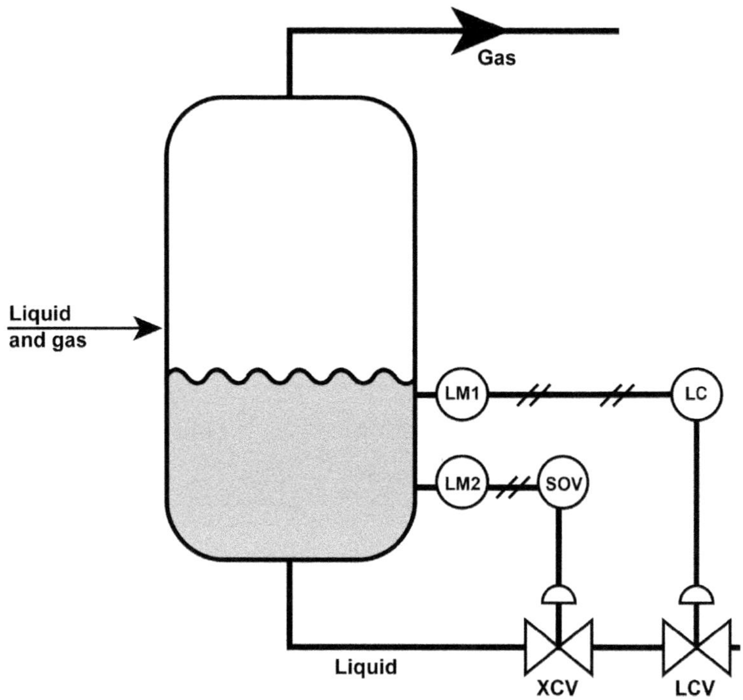


**FIG. 13.3** Simplified diagram for a vessel low-level protection. See text below for an explanation of the abbreviations used.

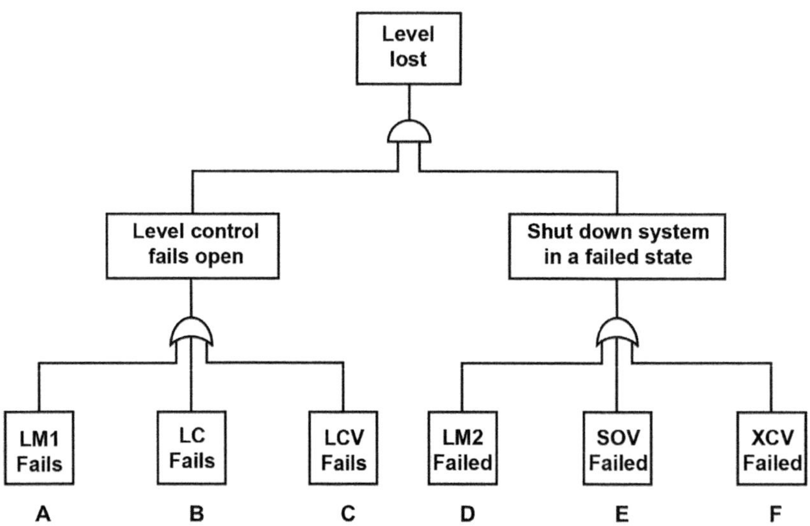


**FIG. 13.4** Developed fault tree for vessel low-level protection.

Common mode failure is likely, indeed probable, when a system contains identical items, relies on common services or has a common maintenance system. In this system LM1 and LM2 are likely to be identical duplicates that could fail in the same way due to dirt in the instrument tappings or because they are tested and set by the same person in the same way. The valves XCV and LCV will have common features, including the air supply and vulnerability to process debris, so a CMF allowance is appropriate. It will be assumed that CMF is possible between LM1 and LM2 and between XCV and LCV.

Thus the entry A in Fig. 13.5, 'LM1 fails', should be replaced by 'LM1 fails non-CMF' and 'LM CMF' with an equivalent change for entry D, 'LM2 fails'. Corresponding changes must be made to entries C and F. A simple revised tree is shown in Fig. 13.6. Note that when the bottom entries are identified care has been taken to assign the same identifier to entries that are identical; thus B and E occur twice. Note that they will have different units in the two groupings but are labelled the same since they derive from a common failure.

Following the same procedure as used in the analysis of Fig. 13.5, the frequency for 'Level control valve fails open' is (A + B + C + D + E) whilst the probability for 'Shutdown system in a failed state' is (F + B + G + H + E). It therefore appears that the combinations of failure that lead to the top event are the 25 combinations of these two groups, i.e. AF, AB, AG, AH, AE, BF, BB, BG, BH, BE, etc. However, when the basic rules of Boolean algebra are applied this number is greatly reduced. Firstly a term such as BB can be replaced by B since the event only has to occur one for both conditions to be met. Similarly EE can be replaced by E. Secondly the term B includes the necessary conditions for any other pair containing B. Thus BF, BG, BH, BE, AB, CB, DB and EB can all be eliminated. After the same

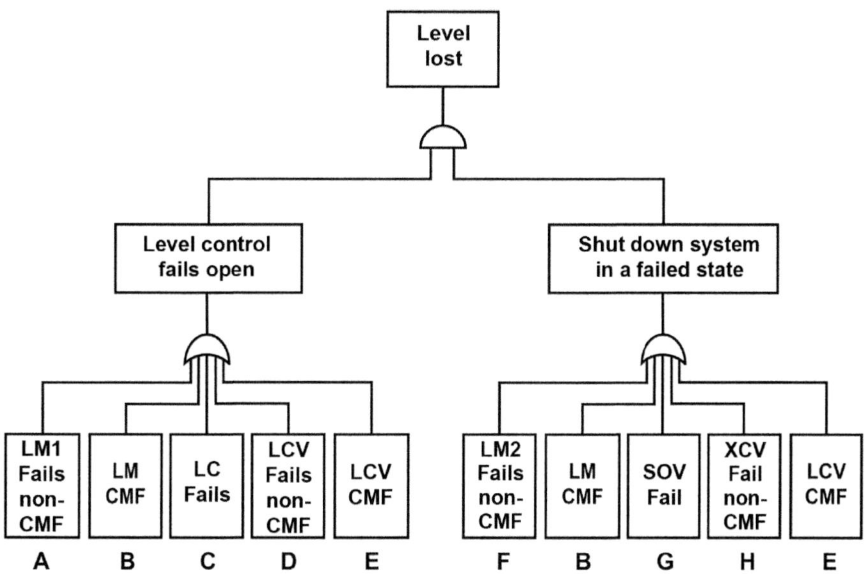


**FIG. 13.5** Developed fault tree for vessel low-level protection, including CMF.

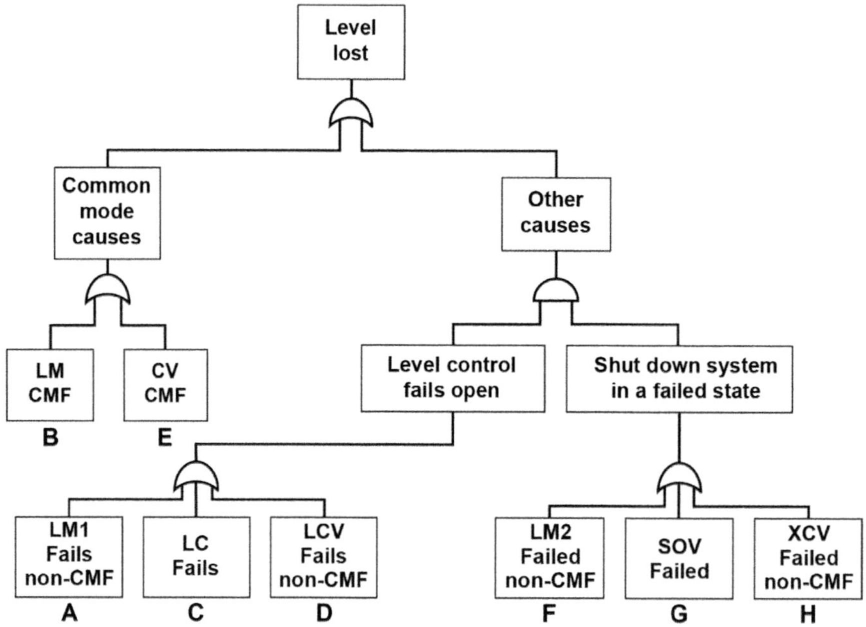


**FIG. 13.6** Final fault tree.

procedure has been use in respect of E the remaining combinations are B, E, AF, AG, AH, CF, CG, CH, DF, DG and DH. These are the minimum cut set for this fault tree. They are expressed in the fault tree in Fig. 13.6, which clearly shows the minimum set of combinations of events, which can lead to the top event.

Final fault tree with all common modes and initiating events drawn out.

Finally, to illustrate the effect on the estimated frequency of the top event, it will be assumed that 10% of the failure rate of the level measurement and the control valve is attributable to common mode effects. This means the CMF rate is 0.01 $\text{year}^{-1}$, leaving the non-CMF rate as 0.09 $\text{year}^{-1}$. The corresponding probabilities are 0.0005 and 0.0045. Using these figures gives an estimated frequency for the top event of 0.0239 $\text{year}^{-1}$ or, on average, once in 42 years. This is very different from the first estimate and shows the need for expert construction and analysis of fault trees.

## Checklists for review of consultants results

Fault trees and event trees

- Did all of the gates have consistent units (probability or frequency)?
- Was the failure rate data use appropriate to that study?
- Were the conditional probabilities appropriate to that study?
- How was the human factor input handled?
- How was the CMF & CCF handled?
- How will human failures be handled?

## References

[1] F. Lees, in: S. Mannan (Ed.), Loss Prevention in the Process Industries: Hazard Identification, Assessment and Control, fourth ed., Butterworth-Heinemann, 2012.

[2] CCPS, Guidelines for Hazard Evaluation Procedures, third ed., John Wiley & Sons, 2011.

[3] G. Wells, Hazard identification and risk assessment, IChemE, Rugby, 1996.

# 14

# Event tree analysis

**Chapter outline**



## 14.1 Definition

An event tree is a graphical representation of the possible outcomes of an incident that results from a selected initiating event. The analysis considers the responses of operators and of safety systems to the initiating event as well as random effects such as the chance of ignition of a flammable mixture when determining the potential outcomes. It can be used qualitatively to determine the range of outcomes and quantitatively to determine their probabilities or frequencies. Event trees are often used with qualitative assessment of the effects in a qualified risk assessment.

## 14.2 Description

Event tree analysis (ETA) starts with a selected initiating event, perhaps a failure. The factors that determine the outcomes are considered in turn. These may be the safety functions built into the system, alternative external conditions which affect the outcome or different operator responses. Each is expressed as one or two alternatives which can be represented as yes/success (A) no/failure ($\bar{A}$). The event tree is constructed, working from left to right, starting with the initiating event and then branching as each factor is considered in turn. Fig. 14.1 shows an event tree where the outcomes are

A Guide to Hazard Identification Methods. https://doi.org/10.1016/B978-0-12-819543-7.00014-8

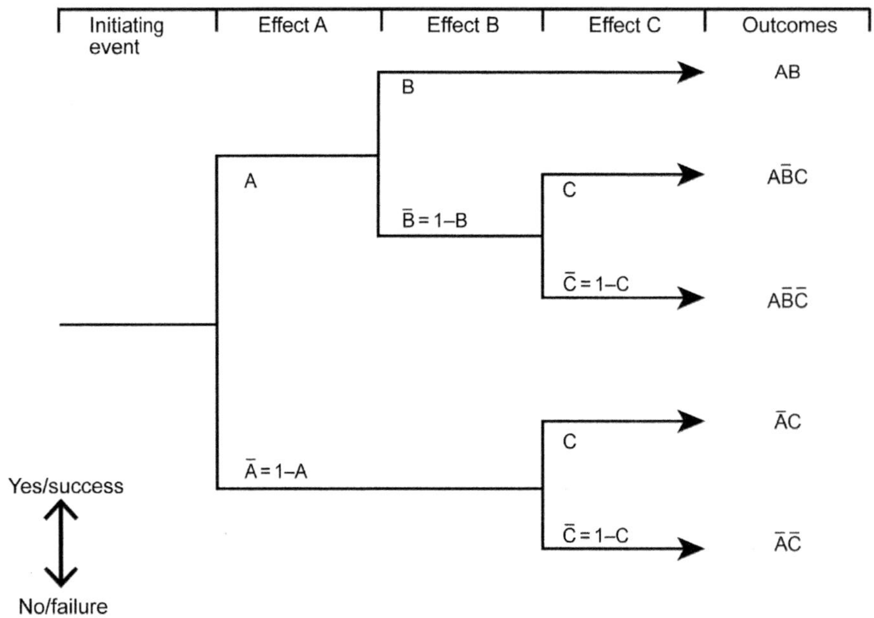

**FIG. 14.1** Simplified generic event tree.

influenced by three factors, A, B and C. Factor B only influences the outcome if A has occurred, whilst C does not affect the outcome is both A and B occur.

Probabilities can be assigned at each branch; they must total 1 in every case. The probabilities of the various outcomes can then be calculated by simple multiplication. These probabilities can be converted to frequencies provided the frequency of the initiating event in known. Event tree analysis is often the partner of fault tree analysis where the frequency of the initiating event may have been estimated.

The outcomes in one event tree may range from a minor to a major consequence. For example, the outcomes from a large gas leak could include dispersal without ignition, immediate ignition and local burning and delayed ignition with explosion followed by continuing fire at the source. It is common to group the end points into a crude classification to give a better overview of the total results (for example on-site injuries or deaths, off-site injuries or deaths, major catastrophes with multiple deaths. (Likewise environmental impacts and nuisance or public reaction.)

The main usage of event trees is not to identify hazards but to analyse the possible history of an event showing all the potential escalating and mitigating circumstances. It gives a short and simple description of all the possible outcomes and provides the means to estimate their frequency/probability. It is simpler to use than Fault Trees and is better suited to the assessment of larger rather than smaller events.

Further descriptions and examples are given in Refs. [1–3].

## 14.3 Resources required

*Manpower*: One person usually carries out the development of an event tree. The skills required are of a lower level than for fault tree analysis, but the analyst must have a clear understanding of the potential contributions to the final event. After completion, the tree should be subjected to independent review.

*Software*: Although there is no need for sophisticated computer aids, there are some programs to assist in the construction of event trees and their quantification.

*Data*: Historical data for probabilities of different outcomes or human actions are required. The comments made in the section of fault tree analysis apply equally here. The data must be robust and relevant. A sensitivity analysis should be carried out when the data are uncertain. [See Risk Assessment (Chapter 15).]

## 14.4 Timing

Event tree analysis is usually carried out early in the project development, often in conjunction with fault trees or other given top events.

## 14.5 Advantages, disadvantages and uncertainties

### Advantages

- The event tree provides a clear and logical graphical description of the potential outcomes from the initiating event.
- The technique is structured and methodical, yet relatively simple.
- Both qualitative and quantitative results are obtained.
- It is well suited to the analysis of complex processes having several layers of safety systems in place to respond to a specific initiating event.
- The analysis confines itself to probabilities that are easily manipulated.

### Disadvantages

- Because of the intensive nature of the analysis, it is usually limited to the identification of special hazards.
- The analyst must be well trained and experienced to ensure an effective study. Skill is needed to determine the degree of detail to be included.
- The probabilities used in the quantification must be robust and relevant.

### Uncertainties

- A critical attitude must be taken to probability data since these are generally less well based than frequency data. Whereas failure rate data have some basis probabilities may be subjective and based on individual case studies. An illustration is the difficulty of stating a probability for ignition of a release of a flammable vapour.

### Applications

- The most appropriate use of ETA is the description of the potential outcomes following an initial event such as the release of material of the failure of a primary protective system where other mitigating features exist.
- Event trees are a valuable contribution to the numerical assessment of risk.

## 14.6 Availability assessment

Event trees can be used as a means of determining the availability of a plant or process. The equipment has two states—available = A, or failed = F. $A+F=1$.

For two systems in parallel the states are: $A^2+2AF+F^2$.
For three systems in parallel the states are: $A^3+3A^2F+3AF^2+F^3$.
And hence the Pascal triangle.

Fig. 14.2 is a reliability block diagram (RBD) for a single, double and triple unit. Obviously the first availability is A.

The second for 1 out of two is $A^2+2AF$ (or $1-F^2$) OR $A^2$ is two out of two.
The third for 2 out of 3 is $A^3+3A^2F$ or $A^3$ for three out of three.

Note that the two out of three arrangement has (nearly) three times the unavailability of a one out of two system.

All of the availabilities are multiplied together. However remember that there are major turn-arounds (maintenance shut downs) that must also be included!

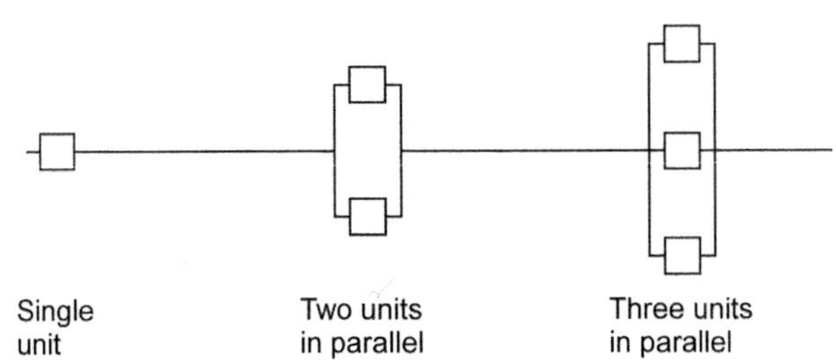


**FIG. 14.2** Event tree for reactor study.

## 14.7 Example

A simple example would be an exothermic reaction runaway due to loss of cooling water to a vessel where a controlled exothermic reaction is carried out by the steady addition of one reactant to the bulk second reactant. The first question in the analysis is whether the stage has progressed so far that, even with loss of cooling, only a small temperature excursion will occur. If this is not the case the first layer of protection is a high-temperature alarm ($TA^{Hi}$) to alert the operator to the developing condition. The operator may then succeed in restoring normal conditions and the batch can proceed. If the temperature excursion still continues a high level a trip system is activated, cutting off the supply of the first reactant. Finally, if the excursion continues beyond this, a runaway reaction develops and the final protection against vessel overpressure and possible rupture, is a bursting disc relieving to a dump tank.

The event tree for these events is shown in Fig. 14.3. It shows four possible outcomes, 1–4 below. With each of these it is possible that the operator is aware of the development (A) or is unaware (U).

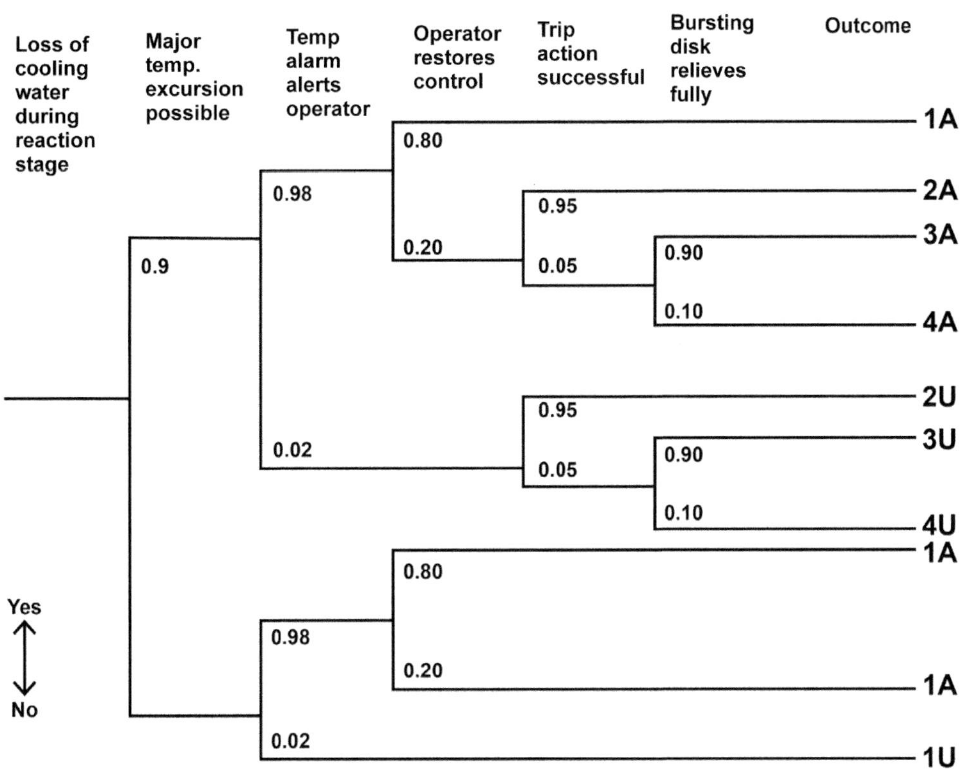


**FIG. 14.3** Simple reliability block diagram.

1. Minor temperature excursion. Batch continues.
2. Large temperature excursion. Batch stopped by trip system.
3. Runaway reaction relieved to dump tank. Batch lost.
4. Runaway reaction leading to overpressure in vessel.

The event tree shows that some of these may be reached in multiple ways. By assigning probabilities at each branch the probabilities of the different outcomes may be estimated. Thus using the arbitrarily assigned values in Fig. 14.3 it can be shown that the probability of outcome 4 is $9.7 \times 10^{-4}$, of which almost 1 in 10 will occur without prior warning to the operator.

It will be noted that event 4U is the most serious, as it is unrevealed, having a probability of ($0.9 \times 0.02 \times 0.05 \times 0.1 = 9 \times 10^{-5}$). This in turn is used to moderate the initiating event frequency.

## References

[1] F. Lees, S. Mannan (Ed.), Loss Prevention in the Process Industries: Hazard Identification, Assessment and Control, fourth ed., Butterworth-Heinemann, 2012.

[2] CCPS, Guidelines for Hazard Evaluation Procedures, third ed., Wiley, 2008.

[3] G. Wells, Hazard Identification and Risk Assessment, IChemE, Rugby, 1996.

15

# Risk assessment

## Chapter outline



## 15.1 Reason for and background to this chapter

It is important that a company who employs a consultant to carry out a quantified risk assessment (QRA) knows both the level of assessment required and can challenge the results from a position of knowledge. These notes are intended to give a 'client' the basic knowledge to ask the appropriate questions when reviewing the report for accuracy and completeness. In addition this chapter contains potential checklists and prompts that can be used during hazard studies stage 2 (Chapter 3), HAZOP (Chapter 4), task analysis (Chapter 6) and bow tie (Chapter 18).

In addition some features are design and operations based. Both groups should be alert to the potentials in the design and compromises in operations.

Anyone who wishes to develop skills in risk assessment must understand the strengths and weaknesses of computer models as well as the potential pitfalls when using them. This chapter should act as prompts.

A Guide to Hazard Identification Methods. https://doi.org/10.1016/B978-0-12-819543-7.00015-X

The general hierarchy of risk-reducing measures is:

1. Prevention,
2. Detection,
3. Control,
4. Mitigation,
5. Emergency response.

Quantitative risk assessment (QRA) is the quantitative evaluation of the likelihood of undesired events and the likelihood of harm or damage being caused, together with the value judgements made concerning the significance of the results. It is usually summarised by three questions:

1. How big?
2. How often?
3. So what?

**How big** uses models of phenomena such as outflow coupled with fire, explosions and gas (and liquid) dispersion linked to the physiological response of humans and objects to events such as thermal radiation, effects of blast waves. This is dealt with in this chapter.

**How often** uses databases on failure rates of equipment and systems and combines them in fault and event trees (Chapters 13, 14 and this chapter).

**So what** requires a value judgement and a set of criteria. Each company/organisation must define its own risk criteria.

Typical criteria are multifacetted and would include, for example:

1. Life, fatality: on or off site: single or multiple: immediate or delayed
2. Chronic, life-limiting injury
3. Injury
4. Loss of revenue flow
5. Cost of repair or restoration (and 4 above)
6. Environmental risk criteria
7. Cost of environmental clean-up
8. Public reaction (local and national). Two examples could be odorous releases and another is the reaction to an event, which was perceived to be one that might have had an effect on the public.

8 is the most difficult to judge but could be the worst in reality!

Note that delayed fatality (sometimes called euphemistically 'premature death') can result from exposure to harmful materials. Examples are given in vulnerability (Chapter 16).

A final criterion must be cost benefit analysis (CBA). Is the cost disproportionate to the potential benefits? The tools for the Identification of Hazards is to be found in Chapters 4–11, the frequency is to be found in Chapters 13 and 14, the vulnerability is to be found in Chapter 16.

## 15.2 Description

Risk assessment can be at three levels of detail. These are:

1. **Quantitative** where the risk is given both magnitude and frequency using tool such as fault tree and event tree analysis and phenomenology. The error bands are about a half order of magnitude.
2. **Semi-quantitative** where the risk both magnitude and frequency is derived from predetermined data sets. Typically this is layer of protection analysis and ranking tools. The error bands can be up to an order of magnitude.
3. **Qualitative** where the risk is based on experience and judgement. The error bands are relatively low as the risks, by definition, will be relatively low. Qualitative assessment should not be used for high-risk events.

While it is easy to understand the magnitude of the effects of an event it is less easy to understand the concept of frequency. Just over 10,000 years ago the north of Europe was covered by an ice sheet. This is about the limit frequency for a fatality.

### Definitions

**Hazard** a physical situation with a potential for human injury, damage to property, damage to the environment or some combination of these.

**Risk** The likelihood of a specified undesired event occurring within a specified period or in specified circumstances. It may be either a frequency (the number of specified events occurring in unit time) or a probability (the probability of a specified event following a prior event), depending on circumstances.

**Individual risk** The frequency at which an individual may be expected to sustain a given level of harm from the realisation of specified hazards.

**Societal risk** The relationship between frequency and the number of people suffering from a specified level of harm in a given population from the realisation of specified hazards.

**Safety engineering** (originally known as **loss prevention**) A systematic approach to preventing accidents or minimising their effects. The activities may be associated with financial loss or safety issues.

**Redundancy** The performance of the same function by a number of identical but independent means.

**It is essential that the words risk and hazard are used correctly.**

Before any assessment of risk can be carried out it is important that the fundamentals are fully understood. These have been expanded below but the message is: avoid all sources of loss of containment, prevention is better than cure.

Before any assessment can be carried out it is necessary to identify the event itself by any one or more of the techniques given, such as Hazard Studies (Chapter 3), HAZOP (Chapter 4) or another. Having identified the event (sometimes called 'The Top Event')

it is necessary to apply the three questions above. The routes to this will probably be quite well defined. Sometimes the event is not predicted because there are poor control procedures in place, lack of critical analysis or lack of basic knowledge. Simple examples are the frequency of significant fires when a building is being refurbished (for example Notre Dame, Paris). More recently there has been the collapse of tailing dams; the first in Baia Mare, Romania, and more recently in Brazil. In each case knowledge was available but it was not actioned properly and not subjected to a critical analysis.
Risks can be thought of in a three-dimensional space as shown below.
They are either:

Chronic or catastrophic. Chronic infers long-term effects and catastrophic immediate effects.
Single or multiple fatalities.
Process or technical. Process is obvious, if the process was changed would the incident be eliminated? The technical is where the fatality may be from a nonprocess-related incident (Fig. 15.1).

The intellectual properties to the ***Safety Cube*** belong to D.S. Scott. The author thanks D.S. Scott for permission to reproduce this figure.

A1 = Single, Chronic, Technical (a broken leg, which does not knit or a damaged eye)
A2 = Single, Catastrophic Technical (nitrogen asphyxiation)
A3 = Single Chronic, Process (gassing or acid burn)
A4 = Single, Catastrophic, Process (small fire)
B1 = Multiple, Chronic, Technical (post traumatic shock)
B2 = Multiple, Catastrophic, Technical (structural collapse)
B3 = Multiple, Process, Chronic (Bhopal or Chernobyl)
B4 = Multiple, Process, Catastrophic (Piper Alpha or Flixborough)

The technical features are more related to Bow Tie (Chapter 18).

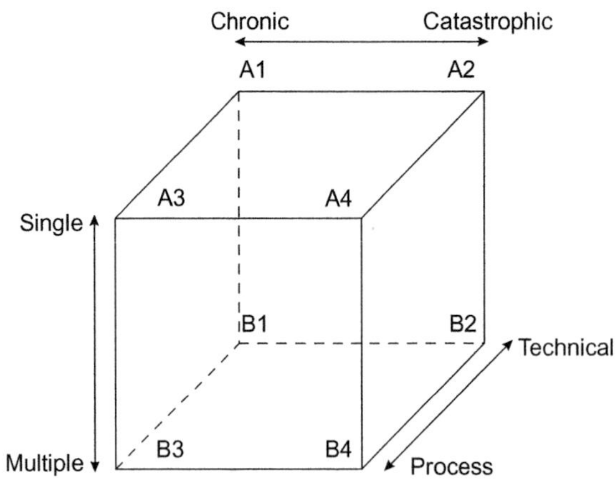


**FIG. 15.1** The Safety Cube.

## 15.3 Resource requirements

### Materials

In the case of quantified risk assessment, the following would be required:

1. Relevant data
   - **1.1.** Applicable failure rate data on humans and equipment. Is it appropriate to the study?
   - **1.2.** Historic results from accident databases
   - **1.3.** Historic evidence of testing of materials for safe use or performance to achieve a national performance standard.
   - **1.4.** Applicable conditional probabilities in event trees (including human failure rates).
   - **1.5.** Wind and weather data—direction, speed and stability (probably from the local airport meteorological station.) Remember that local topological features, such as valleys, may affect the distribution. In addition areas near large expanses of water may have totally different conditions (this is particularly important with offshore systems).
2. Computer programs
   - **2.1.** Fault tree development programs.
   - **2.2.** Verified phenomenology modelling programs to be discussed later but where appropriate these may include stress analysis, fracture mechanics, soil tests, underground surveys (old mines being one example) and the ability of structures to withstand the loads.
3. Corporate risk criteria

What is equally important is that both the assessor and the client fully understand the uncertainties within QRA and can challenge the results from a good understanding of the techniques and models.

### Manpower

The risk assessment can be done by one person but it is better done by a minimum of two persons who can discuss the modelling and data sets.

## 15.4 Timing

Risk assessment is a potential continuum through the life of a project or a facility. There are some obvious times:

1. At Front-End Engineering Design, which would lead to the Safety Case
2. At demolition and site restitution
3. At a mini-project for change in a facility
4. As part of a management of change procedure

5. As part of task analysis
6. As part of a structural modification to a building.

In particular where there are novel or complex issues it would be advisable to carry out a preliminary risk assessment during study 1, concept selection, and before study 2 FEED. Should the risk be found to be high during study 2 there is the opportunity to design it out at source (Inherently Safer). If it is missed and corrective actions are required at a later stage, such as SIS with majority voting, these may be costly and create difficulties in operations.

Following a number of modifications it is possible that the original design intent has become compromised. There is a shift by both the Regulator and Asset Owner to complete reviews of Hazard Studiers (Chapter 3) (including HAZOP; Chapter 4) or PHR in the case of low-risk installation (Chapter 19) on a fixed routine. It should not be assumed that these requirements only apply to process plants; other examples are such as:

- Bulk storage
- Mechanical handling
- Water run-off
- Scaffolding
- Excavations (with unknown underground services and impact on emergency services)
- Any routine operation

## 15.5 Advantages, disadvantages and uncertainties

### Advantages

- The quantification exercise may reveal unexpected contributions to the event.
- This may result in the development of alternative solutions (such as the use of inherently safer design).
- Risks can be ranked in a consistent and logical manner.
- Can be used to assess the risk levels on any process.
- Can be used to identify the causative agents (precursors) in that risk (so leading to prevention).
- Can be used to compare different design options.
- Can be used to identify risk reduction options.
- Demonstrates, to the regulator, that any residual risk is 'as low as is reasonably possible' (ALARP).
- Is equally useful in environmental risk assessment.
- A design can be justified and all contributions listed with the actions for mitigation.

Once the first study has been carried out the risk assessment can be upgraded from operational experience and/or the result of any material change in the process using either a full hazard Study (Chapter 3), HAZID (Chapter 5) or if appropriate PHR (Chapter 19).

### Disadvantages

- Requires specialised skills in the manipulation of data and modelling tools.
- Requires relevant failure or reliability data bases.
- Requires sophisticated modelling tools.
- It is sometime necessary to manipulate the modelling tools so as to obtain meaningful answer. (Gas dispersion models do not model a 90-degree change of direction in a single iterative step—it may be necessary to set the exit point at a couple of degrees to the wind direction).
- The first study should be carried out early in the project development so as to optimise risk reduction. Further if not identified early there is a serious probability that any required changes after sanction is obtained will create a cost overrun and project delay.
- There is a risk that the risk value as derived becomes sacrosanct.
- The rigour hinders lateral thought which then gets ignored.
- The rigour inhibits the use of alternative solutions.

### Uncertainties

- There have to be clearly defined 'risk criteria'. Some may be corporate and others may be defined by the regulator.
- Reliability or failure data (including common mode failure) may be process specific and not universal.
- Human reliability may be part of the corporate culture and training procedures.
- With all risk models the assessor may have to make some assumptions. These must be recorded and justified.
- In the case of gas dispersion a whole series of combinations of outcome may have to be assessed, including direction, wind speed and stability.
- Toxic effects will have to be assessed. This may assume evacuation potential or the duration of release.
- In the case of fires the assessor may have to define the surface emissive power of a fire.
- The assessor will be required to assess the worst-case outcome and probability of that happening.

## 15.6 Understanding the models

### Outflow (source models)

Before any physical models can be analysed it is necessary to know how much (total flow) and the rate at which fluids are produced due to loss of containment. The classic outflow models exist for gas, liquid and evaporation from pools of volatile liquids. Single-phase flow follows the traditional models, but flashing flow requires a specific treatment—Homogenous Equilibrium Method (HEM). In specific cases flashing fluids do create aerosols such that all of the fluid is totally vaporised at source by contact with air. This must be built into the analysis. Equally evaporation from cryogenic pools needs data on the fluid properties as well as the properties of the surface on which it falls. The evaporation rate may change as the surface cools.

In terms of hazard identification the examination of the failure characteristics of equipment and historic failure rate data will go a long way to identifying the areas of vulnerability. Table 15.1 gives some possible sources of leakage (loss of containment (LOC)).

These areas should be reviewed constantly as '**prevention**' by both design and operations groups.

Many of these could be discussed as a Check List (Chapter 10) or in a HAZOP (Chapter 4). In particular standards of isolation and preparation are discussed in Task Analysis (Chapter 6 particularly 'hierarchical task analysis'.)

The solutions are varied. Piping corrosion and erosion can be monitored by non-destructive techniques; this should be followed up through Audits (Chapter 18). Pump seals can be double mechanical seal with an interspace monitor, secondary seals or use of 'seal less pumps'. Jointing subject to chemical attack by the fluids handled can utilise spirally wound gaskets with chemically resistant filler.

**Table 15.1** Possible sources of loss of containment.

| Site of leakage | Possible cause of leakage |
|---|---|
| Piping | Corrosion; erosion, fatigue, metallurgical failure such as stress corrosion cracking, physical over pressure, physical over load due to lack of adequate supports, impact by swinging/dropped loads, people walking on the piping. |
| Vessels/equipment | As above |
| Reactors | Poor ventilation of the work place, poor enclosures, internal exothermic reactions leading to an internal explosion. |
| Equipment opened for maintenance | Poor preparation, poor isolation, poor permit to work procedures, poor supervision. |
| Jointing and flanges | Fair wear and tear, attack by internal fluids, incorrect fitting, poor joint alignment, poor bolting, corrosion of the bolting, jointing failure due to attack from the process fluids |
| Vents | Use and abuse or inadvertently opened by being caught on coveralls.<br>Vents should be blanked off when not in use. |
| Drains | As above |
| Pump seals | Wear and tear, failure of the seal or radial bearing collapse. |
| Compressor seals | As above or seal oil failure. |
| Filters | Use, including rupture of the filter medium, poor procedures, poor isolations. |
| Sample points | Sample point left open or purged into the work area and not a safe disposal point. |
| Maintenance | Poor preparation, poor standards of isolation, poor sweetening procedures. (See task analysis; Chapter 6).<br>Poor restoration of safety systems such as pressure relief valves and emergency shut-down systems |
| Tank breathers | Poor location; process upset and volatile fluids |
| Drum filling | Poor ventilation and controls |
| Analyser houses | Inherent with analysers, poor ventilation |

## Gas dispersion

**Gas dispersion** is a well-developed and understood phenomenon.

### *General introduction and background*

Most dispersion studies will follow the general guide rules explained as follows.

Dispersion is fundamental to the safety of persons both on and off a site. It is necessary to understand what conditions may increase the risk following a release of toxic or flammable gases. It is also necessary to know how far a release of odoriferous gas may travel and still be smelled. The public have a right to a clean and odour-free environment.

The plume below (Photo 15.1) shows the history of a plume of heavy gas released (density about 1.9 that of air) from a vent stack. The vertical section of the plume is the momentum exchange where air is mixed with the gas through vortices at the edge of the plume. The section following is bulk mixing (sometimes called translation mixing) caused by two gas streams travelling in different directions to each other—one is injected into the other in a shearing or smearing action. The third part (as loops) is turbulent mixing due to the local vortices/turbulence within the air stream. The final effect (and by far the smallest) is molecular diffusion or the molecular velocities, which are random in direction. Diffusion is a potential killer as pockets of gas may be left in dead legs in equipment.

**PHOTO 15.1** A plume of a continuous gas release.
**Note**:
(1) Loops this affects the peak concentration.
(2) The plume fall, which is affected by the density.
*Supplied by courtesy DNV GL Spadeadam Research & Testing.*

The history of the plume, and particularly the ground level concentration, is therefore influenced by:

1. Elevation of the release point above the ground level
2. Jet velocity (outflow divided by the cross sectional area of the vent)
3. The outflow in kg/s
4. The angle of the jet with respect to the air flow
5. The density of the gas (MW and temperature)
6. The internal turbulence in the air stream (atmospheric stability and dispersion coefficients)
7. Roughness of the ground surface—sea, hedges or cities all affect the turbulence. Cities are notionally rough but water is notionally smooth.
8. The location of the point of interest (*x*,*y*,*z*) relative to the release point.

The density means that any gas, which has a density greater than air, will reach ground level somewhere along its path. However a lighter gas will tend to rise but some point the extremes of that plume may still reach ground level somewhere along its path. This means that odour and toxic properties cannot be ignored for notionally lighter than air gases.

It is reasonable to note that anything which slows up air flow and creates stagnant zones is a hindrance to dispersion, low wind speeds and stagnant pockets are a potential hazard in themselves. An open, uncongested process plant is safer than a congested plant or a confined space, which has to be force ventilated. A plant where equipment is well spread out is safer than one where equipment is close together. It is recognised that such a process involves more piping and higher capital costs. A plant where there are walls or enclosures are to be resisted as are artificial enclosures such as created by pipe tracks, these all result in a plant, which is less than an open. Ventilation rates from fans are often based on 6–12 changes per hour, which produces 'wind' speeds of about 0.1–0.2 m/s, but even on a very still day air speeds of 0.5 m/s are readily achieved in the open air.

Computer programs, of which there are many, can now give a reasonably accurate history of any plume, but it is recognised that two different models will produce minor differences in the results. The results are usually given as 'time weighted averages'. However the loops evident in the previous photo show that at any point along the plume the actual concentration is varying with time. The actual concentration at any point at time t may vary be ±3 times the time-weighted average.

**This is very important when assessing the likely concentration of an odoriferous, toxic or flammable gas.**

Heavy gas dispersion is a specialised study. LNG (say liquid methane and also ammonia) requires heat from the ground, and more likely the air if released in the form of an aerosol, to complete the evaporation process and at the point in the heat balance where all of the liquid is fully vaporised the final air temperature in the case of methane is −160°C or 113 K. Even though methane has a density of 0.71 kg/m$^3$ at atmospheric temperature compared to 1.22 kg/m$^3$ for air, and while there may be 25% (v/v) methane in the resultant cloud the true cloud density for methane/air at 113 K will be twice that of air. It will sink, not rise, and it will flow as a thin cloud slumping under the effects of gravitational forces on the cloud. (This can be visualised as a spill of water onto the floor). There are methodologies for heavy gas dispersion but they are complex and may be required for other liquids, which are inherently heavier than air such as propane or chlorine.

### *Meteorology*

The atmosphere in contact with the earth is in fact a boundary layer and subject to both temperature and velocity gradients. If a small cylinder of gas was raised from the surface of the earth it would expand adiabatically and cool at a temperature gradient of about −1°C/ 100 m, this is called the adiabatic lapse rate and results in no rising or falling thermals. (In reality it is nearer 0.8°C but the value of −1°C is easier to remember.)

The velocity gradient may not be significant for a small release but is typified by the wind speed at the top of a mountain being significantly higher than at sea level.

### *Stabilities*

Pasquill A stability equate to a hot bright summer day, temperature gradient over −2°C per 100 m. There are strong rising thermals which generate turbulence.

Pasquill F stability equates a cold star lit winter cold/frosty night, the temperature gradient is 0 or maybe 1°C per 100 m. There are descending thermals which create little turbulence and can trap gases in pockets leading to fogs or smogs.

Between A and F there are four levels—the most common being D, the temperature gradient is −1°C per 100 m with no rising or falling thermals. This stability level occurs in the United Kingdom for almost 80% of the time.

There will be some mutually exclusive combinations of wind speed and atmospheric stability. High wind speeds tend to eliminate stabilities A&F. However there will be many potential combinations of wind direction, speed and stability with combinations as high as 50 combinations or more. Each combination has to be tested individually.

In simple terms the very worst weather conditions are with a low wind speed and low internal turbulence (high stability).

### *Basic turbulent dispersion model*

The temperature gradient has been discussed earlier and is of importance and is defined by Pasquill 'Stability Levels'. These feed into the standard gas turbulent dispersion model defined by:

$$x = \frac{1}{K_1} \frac{Q}{\pi \sigma_y \sigma_z \overline{u}}$$

This is the simplified centre line concentration of a plume. Off centre concentrations are defined by:

$$\chi_{xyz} = \frac{Q}{\pi \sigma_y \sigma_z u} \exp\left(-\frac{1}{2}\left(\frac{y^2}{\sigma_y^2} + \frac{z^2}{\sigma_z^2}\right)\right)$$

$\chi_{xyz}$ = The concentration (kg/m$^3$)
$Q$ = Release rate (kg/s)

$\sigma_y$ and $\sigma_z$ are the dispersion coefficients in the $y$ (horizontal) and $z$ (vertical axes). These are a function of the downwind distances as defined by Pasquill and typically have a form of: Constant $\times$ (downwind distance)$^n$. Where $n$ for most stabilities other than A is usually less than 1. See stability below.

$K_1 = 1$ for ground level release and $=2$ for an elevated release
$u =$ wind speed

Obviously the centre line concentration is the highest but the off centre concentration may be of interest.

This equation produces a 'tear drop' shape with the widest point 'about' 2/3 along the path length.

*Inversions*

The rising thermals induced by the sun are therefore a beneficial mixing process. In some cases there are inversions, that is, the density of the air above the inversion layer is lower than that below the layer, so, some pollutants cannot penetrate it by buoyancy alone—and the gases are 'trapped'. (Hydrogen will still rise through the layer as it has inherent buoyancy). Low concentrations of noxious gases such as sulphur and nitrogen dioxides can become trapped with a significant impact on pollution. (These were called 'smogs'.)

- Beware the cold frosty night!
- Remember to make allowance for the time-varying concentration at any point $x,y,z$

*Safe havens*

In some area it may be necessary to install 'safe havens' or 'gas refuges'. These will be notionally air tight rooms. Unfortunately they cannot be totally air tight and there will be some air ingress via air dampers. It might be appropriate to use the time-weighted average concentration to assess the concentration using a continuously back stirred reactor model. Consideration should be given to extra sealing arrangement (such as tape to be fitted round doors and windows).

- If a haven is used it would be appropriate to ensure that there are visible wind socks to determine the safe escape route.
- There should be communication from inside the haven to outside the haven.
- Beware of the rate of rise of the temperature in the haven, each person will release 'about 0.2 kW'.
- Beware of the rate of rise of carbon dioxide in the haven; the exhaled air contains 'about 4% (v/v) carbon dioxide'.

The last two can lead to panic. High carbon dioxide concentrations increase the breathing rate. High temperatures will lead to sweating (perspiration).

## Fires

### *General introduction and background*

Most flame and radiation models will follow the general guide rules explained as follows.

Fires are one of the major process causes, if not the major cause, of loss of production and life on a process plant. Yet the assessment of fires and their effects is still a somewhat inexact science (Photo 15.2).

Fires have a potential for major damage. Steel loses **most** of its structural strength at about 450°C and softens. Equipment handling materials under pressure may rupture due to the reduction in tensile strength and spill their contents into the flame (Boiling Liquid Expanding Vapour Explosion; BLEVE), structures may collapse and joints in piping systems may spring, which then spill their contents into the flame. This is the 'domino' effect.

The combustion of fluids **may** burn to relatively non-toxic gases; however, carbon monoxide is not safe nor is carbon dioxide (an asphyxiate) or carbon particulates (which clog the alveoli in the lung). Compounds, which contain nitrogen or halogens within the molecule and the mix of compounds in a warehouse, **may** burn together to produce toxic materials such as hydrogen cyanide (HCN), phosgene ($COCl_2$) and hydrochloric acid (HCl). This means that downwind there may be hot combustion products, soot and toxic gases.

**PHOTO 15.2** A tank fire.
**Note**:
(1) The downwind lean in a wind of about 1–2 m/s wind. The smoky nature of the Fire – smoke = soot particles + CO.
(2) Unburned paint cooled by the stored fluids—possibly diesel oil.
(3) The systematic collapse of the shells.
(4) Minor wind-induced flame 'drag' down the tank beyond the tank edge.
*Supplied by courtesy of DNV GL Spadeadam Research & Testing.*

The history of a fire is influenced by:

1. The diameter (or hydraulic diameter) of the pool. This means the spread of the fire base. (The hydraulic diameter is the diameter of the equivalent circle with an area equal to that of the pool—which may be square if in a bund).
2. The wind speed (leading to flame lean and downwind drag beyond the flame base leading to different view factors—see later).
3. The chemical nature of the combustible (generation of smoke particles by lack of oxygen or pyrolysis.)
4. The natural aeration by jet entrainment or otherwise of the fire.

*Fuel/air/ignition*

The usual graphical representation of the fire process is in the fire triangle—remember fires may also produce an explosion as a precursor to a fire (Fig. 15.2).

This triangle is a useful beginning for the discussion.

| | |
|---|---|
| **Fuel:** | Fuels are in three phases: solid, liquid and gas. |
| **Solids**: | Could be can dust from wood, coal, metal powders—these can also explode. |
| **Liquids**: | Could be petrol, diesel oil or others. |
| **Gas**: | Could be hydrogen, methane, propane and others. |

In reality it is the gaseous phase that ignites not the solid or liquid phase, so, a wooden log requires some form of 'lighter', just as does diesel dil. (It should be possible to snuff out a match using diesel oil—but please do not try it!)

| | |
|---|---|
| **Oxygen**: | Just as with human life, combustion requires oxygen and flames cannot exist at low concentration of oxygen typically less than 10% (v/v) or a partial pressure of 10 kPa. See flammability envelope later. |

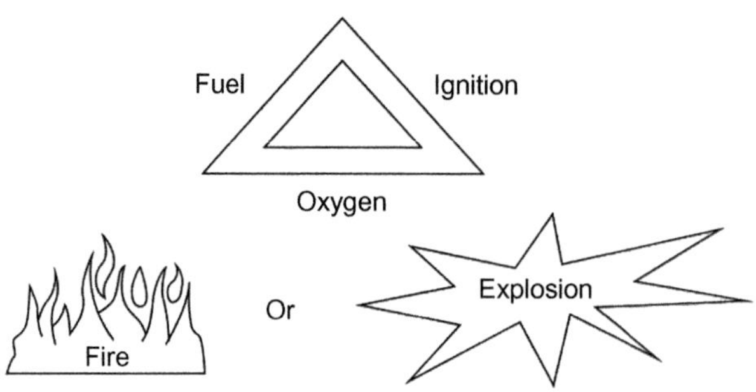


**FIG. 15.2** Fire 'triangle'.

Given the three sides they all have to be together and in the correct balance—take one away—nothing happens.

It may seem odd that metals or dust can explode. A finely divided solid will have a large surface area; these are typified by as corn dust or aluminium powder. If the material is reactive and there is sufficient energy to set off the reaction. Once initiated the exothermic process will take over.

*Flammability envelope*

The fuels are **not** flammable across the whole concentration range of 0%–100% fuel in air. Thankfully, there is only a limited concentration bound in which flames may be supported. This is shown graphically in the flammability diagram or envelope—below. The diagonal line represents concentration locus of a mixture of air and a nitrogen/methane (80% nitrogen and 20% (v/v) methane). At 3 volumes of air to 1 volume of nitrogen/methane the mixture is just flammable. At blends of less than 3:1 with air the blend is non-flammable (too rich) and with blends more than 4:1 the blends are also non-flammable (too lean). Please note that purging out this mixture with air will cut inside the flammable envelope and there may be a point in the purging cycle where there could be a disastrous ignition and explosion. This explains why it is better to purge out the flammable with an inert to a safe location before displacing the residues with air (Fig. 15.3).

This plot has significance when preparing a vessel entry as in task analysis (Chapter 6).

In air (20.8% (v/v) oxygen) the flammable range 'approximates' to 2 times and 0.5 times the stoichiometric concentration. The two extremes where the flame can just exist are the upper flammable limits and lower flammable limits (LFL and UFL) (sometimes 'explosive' is used interchanged with 'flammable'). 'Just Exist' means that the flame will just propagate vertically and only vertically. The 'x 2 and x ½ rule' is not absolute as some gases, particularly hydrogen, do not follow the generalised rule but it is worth remembering if the data are not otherwise available.

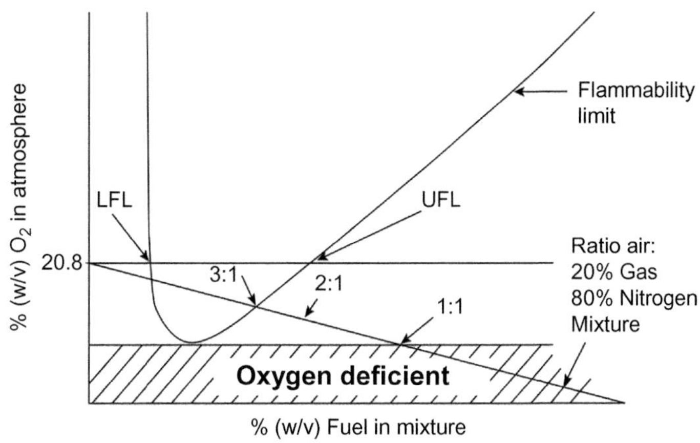


**FIG. 15.3** Flammability envelope for methane.

The envelope widens as the oxygen concentration increases. Ultimately at high oxygen concentrations ignition may occur spontaneously.

As the oxygen level increases the ignition energy falls. High pressure, high temperature and oxygen concentrations all enhance the ignition process and require lower ignition levels as would be expected from simple chemistry and the laws of mass action. Ultimately it is possible for liquids and solids to ignite with ultra high oxygen concentrations in the supporting atmosphere. This has resulted in disastrous fires. The fire in the NASA Apollo modules (Apollo 1, 27th January 1967) on ground test was initiated by an electrical fault but the high oxygen concentrations resulted in a short but intense fire involving plastics, which might not have occurred with lower oxygen concentrations.

Diesel oil has a very low vapour pressure—that is the fuel/air mixture at room temperature is 'too lean' or below the LFL as the flash point of diesel is well above 20°C. Once the oil is warmed locally above its 'flash point' it will ignite. (The flash point is that temperature where the vapour pressure is JUST sufficient to provide a flammable vapour in air.) The flash point of petrol is less and 0°C (or your car would not start on a cold day). *Note*: 'flash point' is a physical ignitability test carried out on potentially flammable fluids using a specially designed piece of apparatus—it has nothing to do with 'flashing fluids'.

### *Ignition energy*

The ignition energy for gases range between 0.019 milliJoules (mJ) for hydrogen to 0.3 mJ for methane. 0.25 mJ may or may not seem much energy—it is equivalent to dropping 1 kg bag of sugar about 25 cm; the energy in the bag of sugar is present but not the intensity so ignition usually requires high temperature localised energy or high intensity. Taken that the composition has to be correct, the ignition energy has to be correct and at the correct location, it is not surprising and fortunately only a small percentage of all leaks actually ignite. Tables are available for the assessment of probability.

### *Sources of ignition*

These areas should be reviewed constantly as '**prevention**' by both design and operations groups.

These should be examined by the check list approach (Chapter 10) or during the HAZOP (Chapter 4). Ask the question 'Can these conditions occur and what should be done about it?'

There are many causes of ignition. The more common ones are formulated below:

**Auto ignition**

The ignition of a fuel by contact with hot materials such as superheated stream mains, hot bearings, welded metal or the like. The values are to be found in the literature but usually are over 200°C and many over 350°C. The quoted values are 'ideals' and just the effect of a wind cooling the surface or a layer of rust or dirt may mean that the actual surface temperatures are lower than anticipated. Do not rely on wind or dirt, but it is fair to note that auto ignition at the quoted temperatures is difficult to achieve because of this and the experimental technique using small vessels.

### Chemical reaction

Oily rags can burst into flames due to slow exothermic reactions. The same is true of oil-soaked lagging, iron sulphide (pyrophoric iron or $FeS_2$) and other reactive substances. The build up is exponential and may take some time to be obvious but once it is warm the rate of temperature rise can be quite dramatic at the end of the process and ignition takes place with no warning.

### Compression

Every diesel car uses compression ignition. Also beware if equipment containing air is pressurised with hydrocarbon (pipelines or pigging equipment) or vice versa. The compression of air and fuels to about 20 bars can cause ignition—so ideally purge out the air with inert gas such as nitrogen (Table 15.2).

### Flame/spark

These include welding, chipping, grinding, electric fault—these are only indicative but give the general clue—do not forget bearings (rolling) or hot drive belts. It is for these reasons that there is a requirement for permits to work and hot work permits.

#### *Permanent electrical equipment*

Fixed, permanent electrical equipment on plants handling potentially flammable materials has to be 'fit for purpose'. This means that the likelihood of flammable materials has to be assessed, the areas identified, the likelihood of the flammable gases being present assessed and finally the appropriate electrical equipment specified and installed. This is handled by 'hazardous area classification' and takes into account the likelihood of the gas being present and the properties (particularly the ignition energy) of that gas.

There are many corporate and industrial codes for this. The following are just three:

- Institute of Petroleum (IP)
- Det Norsk Veritas (DNV)
- American Petroleum Institute (API)

**Table 15.2** Adiabatic compression of air—temp (°C) at P (bar).

| Pressure (bar) | Temperature (max) (°C) |
|---|---|
| 10 | 280 |
| 20 | 379 |
| 30 | 461 |
| 40 | 525 |

**Static electricity**

Static electricity is formed when there is charge separation by two phases moving relative to each other.

- Solid/Solid
- Solid/Liquid
- Solid/Gas
- Liquid/Gas
- Liquid/Liquid

[Gas/Gas is not an apparent source of static].

Static electricity has many causes/sources. When high resistivity fluids are sheared, due to flow or pumping, static can be generated. The threshold resistivity for fluids that might result in the formation of static electricity is about $10^{12}$ ohm cm; this includes diesel oil but tends to exclude fuel oils. Two-phase flow—solids in liquids, flashing flow (two phase flow) or water in oil (mixed phase flow)—enhances the generation of static electricity. Static can accumulate on unearthed lagging, ball valves, filters and even humans. Static can also be caused by splash filling, jetting water, $CO_2$ extinguishers, steam lances, water settling through hydrocarbons and many more. The message is that static is a very potent source of ignition.

Ignition by static electricity probably occurs more frequently than most people believe. When all other sources of ignition have been eliminated it is assumed that the ignition source 'had to be static electricity', but in reality it may have been the main source even when other sources were present.

The two classic forms of static ignition are:

**Lightning**—where liquid as rain drops (or solids as hail) are cycled by strong upward thermals in a cloud. Charges accumulate in the cloud and static electricity, as lightning, discharges between clouds, within the cloud or from cloud to earth.

**Humans**—where solids such as clothing rubs on the body or shoes scuff on carpets (both are solid/solid). In the case of human-induced static it should be noted that it only occurs on dry days or in air-conditioned buildings where the relative humidity is less than about 60%–70%. With higher humidity the charge tends to leak away and cannot accumulate.

The human body has a capacitance of about 200 picofarads (pF) and can be charged to about 10,000 volts. The charge energy ($1/2\ CV^2$) is about 10 mJ, which is more than sufficient to ignite most gas/air mixtures. Another variation is the static generated on a plastic comb following vigorous combing of dry hair (again solid/solid), which can produce small sparks or the charged comb can be used to 'lift' small pieces of paper. The static shock experienced when the human body is earthed is a potential source of ignition. Special footwear may be required.

A variation of this is the filling of a tank with a nonconducting fluid. The fluid itself is charged and only discharges slowly (charge relaxation) to the walls. If the fluid is 'dipped' (to determine its level accurately) using a metal dip tape or rod there could be a local discharge to the tape or rod and a spark causing a fire or worse, a tank explosion. It should be noted that fuel road tankers have a wooden dip rod—not metal—for this reason. (It is worth noting that 'switch filling' tankers are a major source of ignition. In switch filling diesel is put into a tank, which has previously contained petrol. The diesel carries the charge and ignition takes place. Think of the filling of road tankers and of course the 'garden 5 litre containers'!)

Slurry systems, where solids are transported in a turbulent regime on pipes, are even more potent courses of static electricity as there is a complex phase movement:

| |
|---|
| Solid (slurry) and |
| Liquid |
| Liquid and Piping |
| Solid (slurry) and Piping |

**Splash filling**

Splashing organic fluids can generate static changes. It is normal to introduce liquids below the liquid surface to prevent this; filler pipes for petrol/diesel are also electrically earthed through the hose to prevent charge build up.

**Droplets**

Lightning is one obvious example of droplets causing charges.

Water droplets draining through nonconducting fluids can also cause charges to accumulate, so it is essential that care is taken when water settles out of nonconductive fluids, such as wet diesel oil.

Flashing process fluids can produce charged clouds. The cloud then induces a charge on an unearthed conductor, which then results in a small spark where it discharges to an earthed body. Flashing fluids bearing solids as slurries are even more powerful sources of static electricity.

Wet steam from steam lances has been known to ignite organic fluids and under certain circumstances carbon dioxide fire extinguishers can produce sparks from the discharge horn.

### *Other sources of static electricity*

**Grain silos**

Grain is often pneumatically conveyed into storage silos. The silo can also contain grain dust or starch and if the grain is charged there can be a major explosion inside the silo with horrendous consequences.

**Drive belts**

Slipping drive belts between a pulley and a motor or slipping conveyor belts can generate incentive sparks either through friction or static electricity.

**Filters**

Dust filters can become charged during filtration or during discharge of the solids from the filter. If the solids are organic shaking the filter or back blowing the filter can create sparks.

### *Variation on a theme*

Some critical properties of solids and gases are given below. The vapour concentrations of organic vapours are similar to those of solid systems (Tables 15.3 and 15.4).

Also note vapour is given as v/v and the solid wt/volume, the minimum gas concentration in the table above are in the range 0.03–0.05 kg/m$^3$, which interestingly is almost identical to that of solids.

**Table 15.3** Some properties of flammable solids.

| Fuel solid powder form | Lower concentration kg/m$^3$ (g/m$^3$) | Minimum ignition energy (mJ) |
|---|---|---|
| Aluminium | 0.045 (45) | 50 |
| Magnesium | 0.03 (30) | 40 |
| Sugar | 0.045 (45) | 30 |
| Wheat flour | 0.05 (50) | 50 |
| Polystyrene | 0.02 (20) | 15 |

**Table 15.4** Some properties of flammable gases.

| Fuel gas | LFL (v/v%) | UFL (v/v%) | Minimum ignition energy (mJ) |
|---|---|---|---|
| Acetylene | 2.5 | 100 | 0.02 |
| Ammonia | 1.5 | 28 | 40.0 |
| Benzene | 1.4 | 8 | 0.22 |
| Methane | 5 | 15 | 0.29 |
| Ethane | 3 | 12.4 | 0.24 |
| Propane | 2.1 | 9.5 | 0.25 |
| Ethylene | 2.7 | 36 | 0.12 |

The maximum contained overpressure for confined explosions of gases mixed initially at atmospheric pressure is in the range 900 kPa–1 MPa and for solids the range is 500–800 kPa.

### *Heat flux*

Then heat flux at any point is given by the simple rule:

$$\phi = f_{12} \times \text{SEP}$$

Where

$\phi$ is the flux (kW/m$^2$) at the receiving point
$f_{12}$ is the view factor from point 1 to point 2. This requires a three-dimensional analysis, which uses a double integration and is influenced by both the location of the receiver (receiving body) and the wind speed (flame lean and down wind drag).

SEP is the surface emissive power of the flame—kW/m$^2$.
As might be expected each flame has its own SEP and follows the radiation law

$$\phi = \varepsilon \sigma T^4$$

Where:

$\phi$ = SEP (at the source surface)
$\varepsilon$ = emissivity
$\sigma$ = Stefan Boltzmann constant
$T$ = absolute temperature of the flame.

Tables are available for SEP. Translucent flames will have a low emissivity but a high temperature and are potentially more damaging to steel. Smokey flames may have a high emissivity (black body) but due to incomplete combustion will be cooler but when very smoky the SEP may be low.

It is important that the SEP of the flame is chosen with care as it is the one potential weak link in the analysis. There are no fixed values as the SEP will be a function of the smoke density (black body factor) and the flame temperature (partial combustion and/or aeration).

Finally the radiation may not be the critical feature. The gas plume from a fire contains products of combustion, which include:

- Soot (particulates)
- Carbon monoxide
- Hydrogen cyanide
- Hydrogen chloride
- Oxides of sulphur
- Hot gases, which can damage the lung

Flash fires torch fires and boiling liquid expanding vapour explosions (BLEVE) can be treated as variants of pool fires with different SEP and view factors. In the case of flash fires the duration will be short but contact with the skin and ingestion of hot gases into the lung may be fatal. Torch fires the temperature is usually highest because of the good aeration and the heat fluxes on structures on which they impinge can reach 350 kW/m$^2$ due to radiation and forced convection.

## Explosions

### *General introduction and background*

The modelling of explosions is not simple. There are a number of potential models (Photo 15.3):

- TNT equivalent (an old and original method).
- Multienergy (which requires an understanding of the likely turbulence generators).
- Computational fluid dynamics (this is potentially the most accurate but requires a lot of data input).

Major explosions are relatively rare but when they do occur the damage is usually extensive and can extend beyond the site boundary into the public sector. Therefore their potential is not only on site but also off site. The resultant cost of an explosion is major and includes site restoration, loss of sales and probably litigation if property outside the site is damaged.

**PHOTO 15.3** A gas-phase explosion.
**Note**:
(1) The flame boundary.
(2) The flame colour and lack of smoke/soot. This suggests that the mix was nearly stoichiometric or VERY slightly lean.
*Supplied by courtesy of DNV GL Spadeadam Research & Testing.*

The history of an explosion is influenced by:

1. The actual mass of gas involved in the explosion.
2. Localised confinement or the inability to vent the expanding gas cloud.
3. Localised turbulence generators.
4. The fundamental properties of the combustible—fundamental burning velocity. (Some fuels, particularly hydrogen and acetylene, can detonate.)

The difference between a fire and an explosion is the rate and density at which the energy is released. This simple statement may seem surprising but it is the key to differentiating between explosions and fires. On Piper Alpha in July 1988 about 75 kg of fuel was consumed in less than 0.5 s giving an instantaneous rate of 7500 mJ/s or 7500 MW, the residual fire prior to the riser rupture released heat at about 65 M Watts. In a liquid based fire the energy can only be released if the fuel is supplied either as an atomised jet or by the liquid vaporising from the pool due to heat input from the fire itself; therefore, the flame has limitations set upon it by the ability of the oxygen to diffuse into the combustion process—these flames are called 'diffusion flames'. (These are different from a furnace flame where the oxygen (in the form of air) is injected into the combustion zone and the radiation from the furnace assists in the evaporation of the fuel.)

In the case of a flash fire there may be a significant potential cloud of fuel—some kilograms—but the flame does not have a 'feedback' mechanism so there is no explosion—see later—and the flame may only move at a few metres per second.

It is clear that there are pre-requisites for an explosion. Either there must be some form of confinement such as a compressor house or some form of turbulence such as piping, which acts as a fast feedback in the reaction process. In reality the two occur together in most process area. Fire walls round reactors act as confinement and piping and equipment create turbulence. The explosion on Piper Alpha was caused by confinement, while the explosion at Beek in Holland and Flixborough in England was caused by turbulence plus localised confinement.

The drawing below shows the turbulence (as lines emanating from the rear of pipes) resulting from a rapidly advancing flame front on the right. This turbulence mixes the fuel/air mixture and instead of the flame depending upon diffusion it has turbulence to enhance the combustion process. In addition the turbulence at the flame front mixes in the free radicals from the combustion process, and as a result there is a high feedback resulting in flame acceleration, which eventually runs up to an explosion (Fig. 15.4).

The explosion history is not as simple as might be expected. A typical pressure time curve for a typical military or 'condensed phase explosion' is shown below. The positive pressure phase is a sharp pressure rise and decay, which eventually becomes a negative pressure phase marked 'I'. This can produce some rather odd effects such as glass 'sucked' out of a window or a building moving TOWARDS the explosion source as the positive wave hits the far side and the negative phase hits the near side. Others may be the injection of air into a building during the positive phase and then the roof being lifted off during the passage of the negative phase.

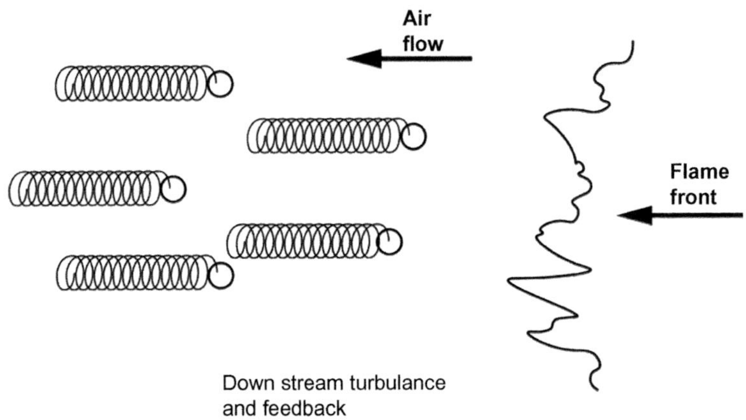


**FIG. 15.4** Turbulent flow regime generated downstream of perturbations.

In a 'vapour cloud explosion' the shape of the positive wave is more like a triangle. However there is still the negative pressure phase.

*Other forms of explosions*

There are other forms of explosion, which should be recognised. They are not an integral part of this chapter but they should be recognised as potential problems. These are given below:

**Bursting vessels**

There are equations for the energy released when vessels burst, this needs some interpretation. Essentially about half of the stored energy is released as blast and half is transferred into flying fragments.

**Rapid-phase transformation**

This is a rapid, almost instantaneous, boiling of superheated fluids. Consider melted metal quenched in say water, the water may become locally superheated (over 100°C) and then suddenly nucleates on metal sites and boils explosively. Similar effects can be obtained by releasing cryogenic fluids under water, the water is the heat source and the fluids boil explosively. The theory is outside the scope of this chapter.

**Compressor houses**

The explosion in a compression house (such as Piper Alpha) may contain from 50 to 250 kg of gases and if it were to be ignited it would result in a large explosion (due to the confinement). In this case they are confined and the overpressure is limited by 'blow out panels'. Such a cloud in the open may cause no more than a 'pop'. Large Vapour Cloud Explosions in open air process plant—now called VCE—may require **1000 kg** or more of fuel before there is an explosion. It is reasonable to assume that this is the threshold and smaller releases are less critical but only in the 'open air'. The analysis of such events would be best handled by CFD.

**Pressure piling**

Pressure piling may occur with a series of chambers and joining corridors, the pressure in the second chamber is precompressed by the explosion formed in first, upstream, chamber. The explosion therefore starts at a higher starting pressure such that if there are a series of chambers the piling may become severe. Eventually walls will fail, so limiting the final effect, but pressures will be larger than expected. Piling can be dismissed from initial studies.

**Explosions in reactors**

In this there is a violent exothermic reaction, which leads to the overpressure of the vessel containing the reaction. This is a complex study, which should be addressed by DIERS (Design Institute for Emergency Relief Systems—AIChE)

**Detonations**

In a gas-phase detonation the flame front travels at a speed greater than the sonic velocity of the unburned gases, velocities may reach 2000 m/s and the overpressures may reach 20–25 bar. The detonation can be achieved in very rare conditions, but it is most unlikely and can be dismissed for initial studies. Suffice it to know that under very rare conditions of turbulence **and** confinement such as pipes and with some gases detonation can occur but it is an extreme event.

**Explosions in houses**

Explosions inside houses have to be modelled in a different manner. Once again the theory is outside the scope of these notes. However pressure piling may result in the destruction of a structure or building resulting in systematic collapse). Typically walls will blow out at a fraction of a bar overpressure.

Strange effects have been notes. The two pressure time plots Figs. 15.5 and 15.6 show both a positive and negative phase. If the positive phase hits the extreme end of a shed at the same time as the negative phase hits the upstream side the shed could move towards the source. Likewise if there is pressure injection in the positive phase followed by the negative phase a roof might appear to be lifted off.

*Explosions in ducts*

Sometimes low-pressure ducts are sealed with a flexible membrane such as metallised canvass. The canvass is ruptured by the internal explosion in the duct and then it is drawn into the duct by the negative phase!

**Military explosives**

It is appropriate to make note of the differences between gas phase explosions and the military use of explosives. The military use 'condensed phase' explosives where the oxidiser is within the fuel. Peak pressures at the front can be thousands of bars. Various formulations can be made to enhance the fragmentation effects. Of course there are special bombs, which carry only the fuel which is then sprayed into the area of attack and then seeded with detonators driving the flame front into a vapour phase detonation.

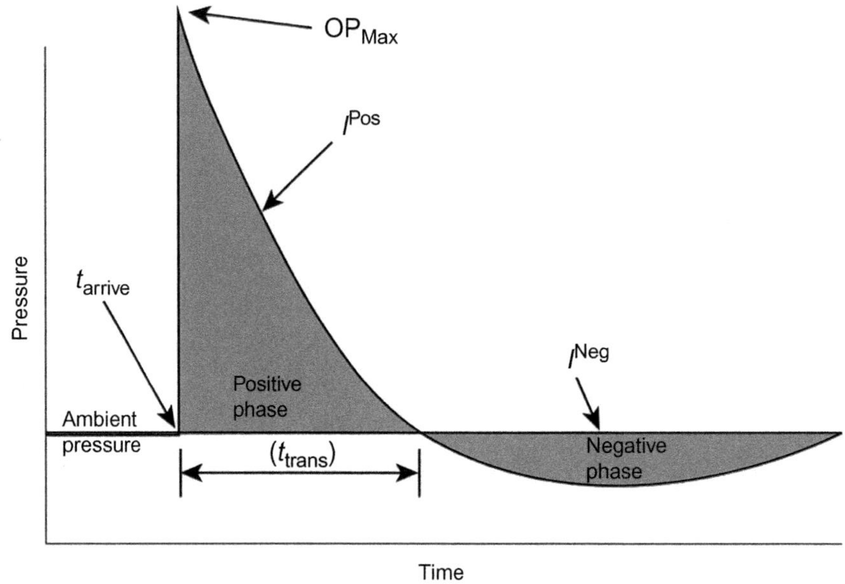


**FIG. 15.5** Time v pressure history for a military explosion. $t_{arrive}$ is the arrival time, $t_{trans}$, OP is the peak overpressure, and $I$ is the +ve or −ve impulse, which may affect the ultimate damage.

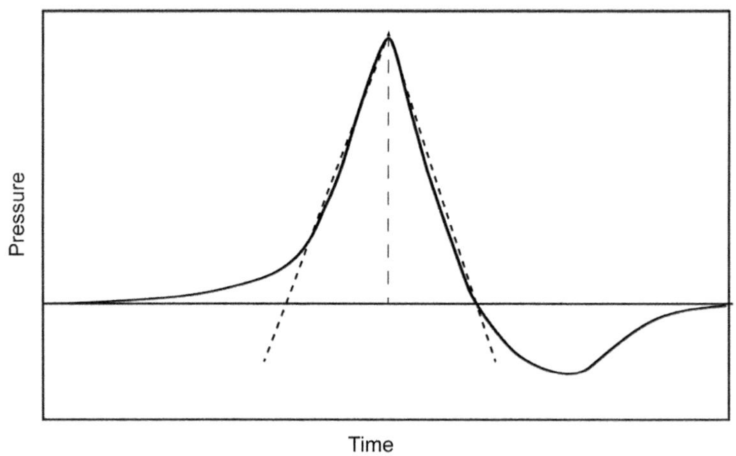


**FIG. 15.6** Pressure v time history of a VCE.

There are a number of models available, but it is important to understand the mechanisms being used and their applicability to the specific case.

It would be wrong not to mention that ex-military explosives are being dug up throughout Europe (and elsewhere) during new developments. Evacuation distances can exceed 500 m.

Tables for resultant overpressure damage in ducts, buildings and the result of imposed overpressures are available in the press. It would be inappropriate to choose one against another.

### Finally

In the case of fully quantified risk assessment the error bands are about a half order of magnitude and cost benefit assessments will be required.

Semi-quantified risk assessment **must not** be used where the risk, as defined by the criteria, is severe. More particularly the use of a 'risk matrix' is not recommended. (See LOPA; Chapter 7.)

The logic flow in LOPA (Fig. 7.3) shows how cost benefit analysis (CBA) can be incorporated to prove 'as low as is reasonably practicable'. The recycle loop can be repeated with new or extra barriers in place and the cost of each measured against the risk reduction.

## 15.7 Checklists for review of consultants or users results

The general hierarchy of risk-reducing measures is:

### Prevention

Loss of containment see Table 15.1 earlier.

Can the materials of construction handle the extremes of process temperatures (high and low)?

Are the materials of construction suitable for any potential corrosive or erosive conditions?

Are jointing materials suitable for the fluids handled? (Some joints are attacked by the fluids handled)

Are only essential joints uses and 'all welded' used wherever possible?

Has consideration been given to the extremes of pressure and the need for Pressure Relief Valves?

Are rupture disc appropriate when there is a rapid pressure rise?

Has the possibility of an internal over pressure (exothermic run away) been considered and the impact assessed? Is there need to use DIERS (Design Institute for Emergency Relief Systems—AIChE)?

Has attention been paid to the pump seal arrangement? Double mechanical seals with interspace pressurisation with a compatible fluid? Secondary seals with interspace flushing with a compatible fluid? Seal-less pumps? (Compressor seals are a special study)

Have the shutdown systems been designed against suitable criteria and can they be tested under conditions that as simulate reality?

Has common mode and cause been assessed? (See Chapter 13; Fault Tree Analysis).

Has 'human reliability' been incorporated correctly?

Are the maintenance and inspection routines appropriate to the process?

Are there safe and secure procedures for recommissioning Pressure Protection and ESD following a shut down?

Is non-destructive testing justified?
Are the isolation standards appropriate to the pressure and nature of the fluids handled?
Are there secure Management of Change Procedures?
Are there secure maintenance and hand over procedures in place?

## Detection

Are all parameters monitored and where necessary independent verification installed?
Do the operating instructions give the operative the correct advice for handling deviations?

## Control

As above

## Mitigation

Is pressure relief required?
Is fire protection required (active or passive)?
Have the shutdown systems been designed against suitable criteria and can they be tested in a realistic manner? Can they be inadvertently isolated from the process?
Has human reliability been incorporated into the analysis?
Has common mode and cause been assessed?
Is the surface suitably graded to prevent accumulation of fluids toxic or flammable?

## Emergency response

Are the off-site response teams fully aware of the materials handled and how to treat loss of containment?

## Leak rate

Was the failure rate in the form of holes size and frequency spectrum case specific?
Was consideration given to common cause effects?
Was the outflow assessed by a methodology which was appropriate to the phase of the fluid?
Was consideration given to possible leaks during maintenance?
Was allowance made for common cause or mode failures?

## Dispersion

See leak rates
Was the weather data appropriate? Did it have sufficient spread in direction, atmospheric stability and wind speeds?

Were there any local topological features, which might have affected the wind direction spread?
Was the source specified correctly (velocity, vent size, height, angle of orientation, fluid properties.)
Were the potential targets clearly specified?
Was surface roughness appropriate?
Were the potential targets clearly specified?
Were sufficient combinations of stability, direction and wind speed and local changes in elevation carried out to specify the absolute risk?
How was the vulnerability assessment carried out?

### Fires

Was there an assessment of all possible wind speeds?
Was there an assessment of all potential wind directions?
Were the potential targets clearly specified?
Was the view factor appropriate to an elevated flame and ground-level receiver (or vice versa?)
Was the SEP appropriate to the flame and fuel source?
How was the vulnerability assessment carried out?

### Explosions

Was due attention made as to confinement and turbulence on the final outcome?
Was due attention paid to the fuel burning characteristics?
Were the potential targets clearly specified?
How was the vulnerability assessment carried out?

## References

[1] F. Lees, In: S. Mannan (Ed.), Loss Prevention in the Process Industries: Hazard Identification, Assessment and Controle, fourth ed., Butterworth-Heinemann, 2012.

[2] There are many examples of Risk Assessment at www.hse.gov/risk/.

[3] The Health and Safety Executive's Five steps to risk assessment. This is in general terms.

## Useful reading

IEC 31010:2019 Risk management—risk assessment techniques.
There are many consultants who can offer this service.

To be selective in the references ignores the fact that each topic needs its own set of references and it is difficult to give each the appropriate breadth of cover.

# 16

# Vulnerability

**Chapter outline**



Perforce this chapter must deviate in format from others. It is designed to interface with other chapters particularly Task Analysis (Chapter 6), LOPA (Chapter 7), RAST (Chapter 9), Quantitative Risk Assessment (Chapter 15) and Bow Tie (Chapter 18). It is not a structured Hazard Identification Method but more one based on some of the more common vulnerabilities of humans and hardware.

Further it is recognised that poor safety management systems, of which there are many, have a potential for human injury. The discussion of SMS does not fit well into this guide and is better covered by audits (Chapter 17).

A Guide to Hazard Identification Methods. https://doi.org/10.1016/B978-0-12-819543-7.00016-1

This chapter will not address the impact on water courses. However it is a timely warning that nitrate and phosphate run off from agricultural systems produces significant 'algal bloom', which can be toxic to humans!

## 16.1 Human vulnerability

### Introduction

Humans and hardware are likely to be subject to assault, injured or damaged, by a number of routes. Each route may be specific to the target object. It is obvious that the prime objective is to prevent such assault routes by designing them out. Before this can be done it is first necessary to recognise what are the assault routes and the effects. The physical assaults are best handled by inherently safe principals and avoidance and only in the last resort the use of physical protection. In the latter case it may be necessary to delineate 'hazardous area' marking them as 'no-entry zones' or requiring the use of Personal Protective Equipment (PPE) for entry into those zones (a SMS in operation). Ultimately in the case of potentially toxic or harmful materials as well as noise it will be necessary to carry out regular health monitoring.

The chemical assaults, in the United Kingdom, are summarised in the **C**ontrol **o**f **S**ubstances **H**azardous to **H**ealth regulations (COSHH). These itemise amongst others:

Chemicals
Products containing chemicals
Fumes
Dusts
Vapours
Mists
Nanotechnology
Gases and asphyxiating gases
Biological agents
Germs (and viruses) that cause diseases.

While not specifically stated others could be irritants and narcotics.

While it is important to look at the chemicals in a process industry COSHH applies also to the general public as has been demonstrated by Legionella pneumophila from cooling towers.

Lead, asbestos and radioactive substances have their own regulations.

### What are the assaults and routes?

The identification of substances hazardous to heath requires and understanding of the properties of the substances in the form of a 'risk assessment'. There are a number of sources for the data:

1. N.I. Sax—Sax's Dangerous Properties of Industrial Materials, 8th ed. by **N.I. Sax** (ISBN: 9780442016753))

2. Bretherick—Handbook of Reactive Chemical Hazards, Academic Press, 2006, ISBN 9780123725639
3. EH40/2005 for work place exposure limits (UK)
4. Safety Data Sheets http://www.hse.gov.uk/coshh/basics/datasheets.htm

The latter is driven by an EU Directive REACH (**R**estriction **E**valuation **A**uthorisation and restriction of **Ch**emicals).

It is important that it is recognised that these values apply not only on site but off site. Off site concentrations are potentially more difficult to handle and are covered by different Regulations/Standards.

## Beware the unexpected

While the reactants and products may be well defined, it is possible that a bi-product or intermediate products may be toxic. In the 1960 there was a process for the production of α-naphthylamine, it produced a by-product β-naphthylamine in small quantities. α was not harmful but β was found to be carcinogenic. The process was abandoned after three unexplained premature deaths. No one anticipated that the by-product might be a carcinogen. (Source—in house knowledge of a company that does not now exist.)

Between 1953 and 1967 mercury was released to the sea at Minimata, Japan. This resulted over 4000 persons being affected, through eating contaminated fish, and an uncertain number of deaths quoted in 'The Guinness Book of Records 2000'—'Worst Marine Pollution as 43 to 800'. It became known as 'It hurts' as joints were affected.

### *Toxic doses*

It is obvious that the simple calculation of concentrations of gases, thermal radiation or overpressures does not tell the whole story. The next step is the 'effect' on the human. (In most cases it will be found that the human is the limiting factor).

The effects are not absolute values but are an analysis of historic data and are based on a typical sample of the population. Individual responses will vary so that effect on one person of different ages or state of health 'challenges' will be different from another and any analysis should bear in mind for example the location old folks homes and schools. (This is built into the Land Use Planning Regulations; Chapter 1.)

Toxic doses are an evolving study, for example the maximum exposure limit (MEL) for benzene has fallen from 25 to 1 ppm over 50 years. Further how long will that incident last?

The effect of high concentrations of carbon monoxide on a piece of process equipment is negligible but for humans it is different. The dispersion (consequence) models will show what is the instantaneous effect but some interpretation is necessary when 'total exposure' is the problem.

The dose models have the form:

$$\text{Dose} = \int C^n t$$

where

$C$ is the concentration of toxic (ppm usually v/v)
$t$ is the exposure in seconds, minutes or hours
$n$ is a constant.

Time is clearly a variable. Further the dispersion models (See Risk Assessment (Chapter 15) give a 'time-weighted average' (TWA) concentration. The analysis must therefore consider:

What is the effect of **that** incident, for **that** time?

This requires an understanding of who will be affected and for how long—the dose. The human physiology is not the same as that of animals. Certain chemicals produce different effects. 'Thalidomide' is the classic example that tests on animals could not replicate the effects on humans. Animal tests although beneficial may not tell the whole truth. The concentrations are fixed for reasons of reproducibility and no allowance is made for spikes or peaks. Tests on animals are supplemented by reference to real events and the assessment of local concentrations.

The standard method for assessing the dose effect is to plot log time vs log concentration and then to group those where the effect was unpleasant, those where the effect was harmful and those where there was some fatality. The slope of the line is then the power in the dose equation (positive or negative depending on how the data are plotted) (Fig. 16.1).

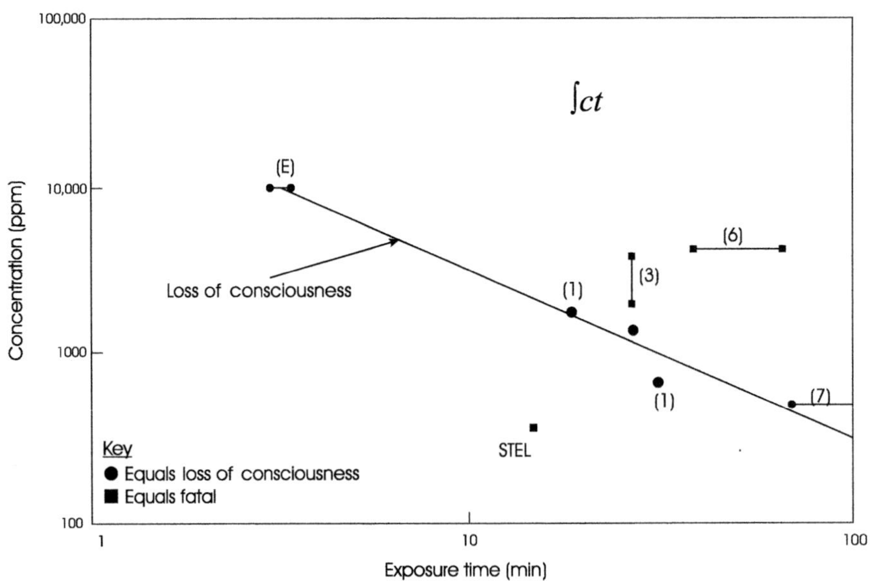


**FIG. 16.1** Carbon monoxide toxicity dose derived by author.
(The accepted slope is now about 0.88.)

The value of $n$ in the dose equation, above, varies from toxic to toxic. For carbon monoxide $n$ = just under 1, for Chlorine and Ammonia $n = 2.75$. (For $NO_X$ or oxides of nitrogen the value of $n$ can be higher). What this means is that doubling the concentration reduces the exposure limit by a factor of 6–8. Of more importance is that the actual concentration at any point in a plume defined by the dispersion models is varying with time by a factor of ±3 so the true value of: Dose = $\int C^n t$ **may** be significantly higher than might be expected from the time-weighted average (TWA) from the dispersion equations. In fact it is possible that the dose derived from the TWA value of C might be classified as only 'harmful' but the true dose, taking into account the peaks, could be fatal. Some toxics are not harmful at low concentration levels (Work Place Exposure Levels), but nitrogen oxides have a persistence, which can result in fatalities from pneumonia some days after the event due to damage to the cilia in the bronchus.

## Other doses

Doses for chemicals are to be found in data books. Solid and liquid toxics are often given as $LD_0$ or $LD_{10}$ or $LD_{50}$, the dose which will produce 0%, 10% or 50% fatality in a group. The values are usually milligrams per kg weight. $LC_{10}$ is not really sensible as it infers that 10% of those affected have been killed.

Other doses are often given as OEL (Occupation Exposure Limits) in tables.

TLV is the threshold value for 8 h per day. STEL is the Short-Term Exposure time for 15 min. **BEWARE** the values are often revised downwards every year as the full effects are reanalysed and assessed.

### *Low oxygen partial pressure*

Once the oxygen partial pressure falls below 10 kPa humans lose consciousness. This applies to confined or trapped spaces. There are a number of examples in the IChemE Loss Prevention Bulletin. The following is one that was witnessed by the author and is illustrated by the following diagram. There were two process fluid circulations. The lower section is co-current over 'disc and doughnut' plates and the top by counter-current over standard trays (Fig. 16.2).

The following dialogue is added for completeness.

The plant was to be inspected during a major overhaul and as part of the preparation the internals of the contactor was 'steamed' by passing hot steam at 150°C for 6 h at point 'A' and then displacing the steam by nitrogen at the end of the 6 h. The nitrogen was displaced by 'the chimney effect' through air entering at the lower manhole (MH) and leaving at the top manhole (MH). The two routes are shown in the diagram.

All process lines were sealed, after the system had cooled, by metal plates known as 'slip plates' (as above) on the contactor side of every valve flange as shown. Air tests for oxygen and flammables were taken at the man holes top and bottom as shown. The oxygen samples at the manholes were 20.8% (v/v) and the flammables were not detectable (N/D).

Note that there is a dead pocket, which could have resulted in a fatality. Fortunately it was identified just in time!

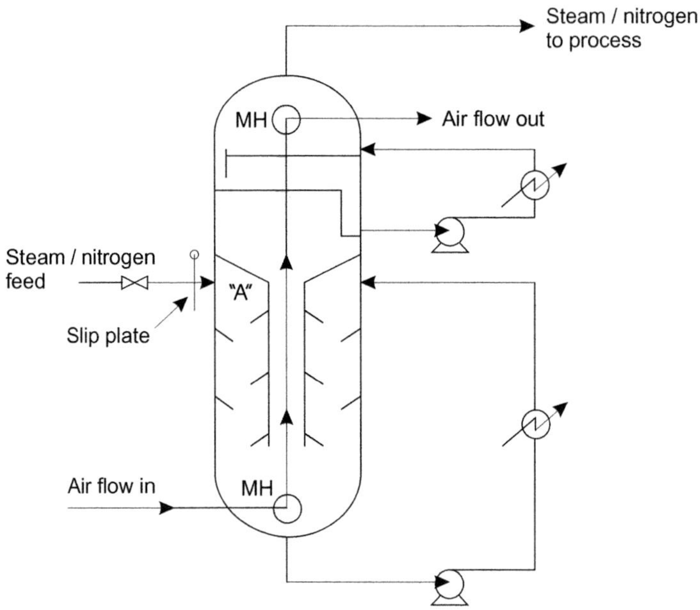


**FIG. 16.2** Gas contactor.

## Other effects

### *Mutagenicity*

Some chemicals can produce mutagens. Nuclear radiation is just one example.

### *Teratatogenicity*

Some chemically can produce deformities of a foetus before birth. Thalidomide is just one example. The tests were carried out on animals where the physiology was different to that of humans and the foetus of was not affected. As a result it was accepted as 'safe' for gravidus humans. There is some evidence that Thalidomide is/was also mutagenic.

One substance can exhibit a number of different effects (glue sniffing causes narcosis, irritation of skin, mouth and nasal passages and systemic damage).

### *Carcinogenic materials*

Many industrial chemicals are capable, to some degree, of causing cancer, and the list of potential carcinogens is growing every year. Exposure to these materials produces effects, which are delayed, typically, 15–20 years. The mortality from the nuclear event at Chernobyl is still evident but the numbers of 'premature deaths' is dropping off rapidly, this may be a result of natural ageing processes. Please look at the earlier section, which notes that toxic levels are generally revised downward with the increase in knowledge.

The handling of some materials presents particular problems since it is generally accepted that there may be no 'threshold dose' below which no damage will be done. Exposure limits are set at levels where risks are acceptable. Benzene is one such chemical

**Table 16.1** Some properties of toxics.

| | |
|---|---|
| Irritants and toxics | Respiratory (chlorine, ammonia, $NO_x$, etc.), skin, eyes. Irritants affect the body tissue at the site of contact. Effects range from discomfort to death. Some make the victims more susceptible to infections such as pneumonia. Panic result from exposure to irritant gases and the response may be unpredictable reaction known as '***fight or flight***'. |
| Narcotics | Some common chemicals, e.g. hydrocarbon vapours produce narcosis, which can interfere with an individual's ability to look after themselves. |
| Asphyxiates | Simple suffocation by reducing oxygen concentration, e.g. nitrogen (see above) |
| Chemical | Some link, in competition with oxygen, in the blood haemoglobin (carbon monoxide, hydrogen cyanide being two) |
| Systemic poisons | These can enter the body through the skin and cause either temporary or permanent damage to the body system. |
| Dusts | Many dusts can lodge in the lungs and eventually produce disease. Historically, three times as many coal miners died from lung disease attributed to dust than died in accidents. |

where the true 'safe threshold' is not known accurately and is falling with time. Hydrazine, aniline and formaldehyde are three other carcinogens (Table 16.1).

## Other harmful materials

Liquid toxics can enter the human body via the skin or the mouth. The skin is not impe-rious and nerve gases can gain entry through this route. There is also the problem of acids and alkalis as well as sensitivity to certain chemicals—even nickel can produce allergic reactions and chromium can produce necrosis of the septum in the nose.

Gaseous toxics may enter the human body via the skin or more likely the lung. Some can result in that person becoming susceptible to lung diseases particularly pneumonia.

Solid toxics enter the human body via the stomach, skin or the lung.

Different toxics have different effects on the body systems.

The toxic data are usually derived by plotting the concentration and exposure duration on a log vs log graph against time with the recorded physiological effect.

Carbon monoxide produces carboxyhaemoglobin, most other toxics affect the lung for example chlorine and ammonia produce pulmonary oedema, while nitrogen dioxide can damage the lung and the cilia and so make the invalid more susceptible to pneumonia. These effects can be delayed for a few weeks!

For carbon monoxide a dose of **about** 4000 ppm for 10 min is likely to be serious.
For chlorine a dose of **about** 75 ppm for 5 min is likely to be serious.
For ammonia a dose of **about** 2500 ppm for 15 min is likely to be serious.
For $NO_X$ a dose of **about** 250 ppm for 10 min is likely to be serious.

### *Physical protection*

See **P**ersonal **P**rotective **E**quipment Regs in UK

'The human' is vulnerable to many physical assaults. The human cannot perform in heat, cold, reduced oxygen concentration, acceleration, dull or bright light and loud noise. Consider the following (Table 16.2):

**Table 16.2** Some sources of human vulnerability.

| Area | Some sources of human vulnerability |
|---|---|
| Head | Impact; loose hair being caught in moving equipment |
| Eyes | Light; low or high; ultra violet light (welding); lasers: grit; acid; alkali; dust; projectiles including weld spatter |
| Ears | Noise—120dBA equals pain, 90dBA for 8h equals hearing loss |
| Nose/lungs | 'Toxics'; dust; low oxygen concentration; hot, dry air; nuclear radiation |
| Skin | Acid; alkali; nuclear radiation; heat; cold; thermal radiation; projectiles; sharp objects; trips/falls, burns by weld spatter |
| Bone | Heavy objects; trips/falls |
| Brain | Information overload/cognitive dissonance/mind set/stress/confusion/lack of training/panic |
| Balance | Wind gusts of 30m/s or acceleration levels of 0.1g laterally will result in loss of balance |
| Muscle and tendon damage | Poor ergonomics: stretching; lifting; trying to exert forces by load oblique paths, etc. |

The list is not complete but is sufficiently detailed to show that the human needs to be protected with care.

Before going into detail it is obvious that persons working at a height will require a body harness and secure eye bolt if there is a strong wind or narrow access.

Likewise when working in confined spaces apart from monitoring the environment it may be appropriate to have 'a stand by man' and harness and lifting frame for emergency exit.

The industrial safety helmet will give reasonable protection against a light impact or a light dropped object from above. (Say 1 kg dropped 10 m.) It will not necessarily protect the wearer against walking into a low beam nor will it protect against a sharp edged or pointed dropped object.

The industrial safety spectacles and full wrap-round goggles give good eye protection against dusts, but it **may** be necessary to use a full-face visor for hazardous fluids. 'Green Glasses' can protect the eyes against intense light when examining a furnace. Special face visors and protective glass is necessary for electric arc welding.

At noise levels over 100dBA communication is difficult and the threshold for noise-induced hearing loss is sometimes quoted as low as 85dBA for 8h. Hearing loss is accumulative and initially speech appears to be distorted, as the high-frequency elements are lost, ultimately there is a problem of 'tinnitus' (hissing or ringing sounds in the head). A 'disco' is a powerful source of damage. If 90dBA is taken as the threshold, 93dBA for 4h is equal to 90dBA for 8h, likewise of 96dBA for 2h or 99dBA for 1h or 102dBA for 0.5h. Ear muffs fitting over the ear with soft seals against the face are the best protection, but noise leakage is still a potential issue. In addition beware of a poor seal (leakage) with a beard! (Ear plugs give some protection but noise can still be transmitted through the human tissue).

Face masks (filters) are available for dusts but they are of no use for harmful gases. (It should be noted that the 'Gas Masks' used in WW2 were of the activated charcoal absorbent type.) The use of Breathing Air (BA) masks is necessary. BA may be by a demand valve when the pressure in the mask falls below atmospheric pressure (as a SCUBA air mask) so leakage can take place around the face mask seals or it may be by a supply valve, which keeps the face mask slightly over pressures with respect to the atmosphere.

Gloves come in various forms. Standard gloves will protect against cold and also can have 'rubber studs' to enhance grip. 'Leather' gloves can protect against 'sharps and rags'. However special gloves, mitts or gauntlets will be needed for hot, cold or harmful duties.

Steel-tipped safety boots, like helmets, can give protection against light-dropped objects. (Again about 1 kg dropped from about 10 m.) Special boots may be needed for harmful fluids.

Industrial 'cover-alls' can give good, general, protection against benign fluids but it may be necessary to use acid/alkali resistant clothing. In some areas it might be appropriate to wear flame-retardant protection.

**(Remember to tuck clothing and such that the spills shed away from the tucks and not into them)**

Dry air at 100°C will damage the mucous membranes in the lung; dusts or smoke can coat the lung (more miners die from silicosis than injury).

The skin cannot tolerate temperatures less than −10°C or temperatures over 60°C for more than a few seconds. Thermal radiation of 6 kW/m$^2$ for 20 s produces real pain. Review the need for heavy, heat resistant, clothing and/or fireproof coveralls.

Consider the impact of 'confusion' and also the dangers of wind load on the body when working in stormy weather.

Consideration should be given to well located 'escape' breathing air sets (BA) round plants handling toxic fluids.

### *Safe refuges*

As the dose is the area under a curve the integral of the ingress of toxic into a 'refuge' or building produces a rapidly worsening situation. Ultimately if the building is fairly leaky the person may be incapacitated and too weak to help in any rescue. Fresh air room must be leak tested and be carefully located. Consider the need for communication outside the room. In addition consideration should be given to strategically placed windsocks. (The rate of toxic ingress could be defined by the TWA values from the dispersion equations.) Refuges must reflect the impact of the increase of carbon dioxide (increased respiratory rate) and increased temperature (general discomfort). Both can result in panic!

### *Probits*

There should be some mention probits for the completeness of 'Toxics'—the task is to **avoid** any risks—**prevention** being the objective.

The dose effects can be converted into probability of fatality by a probit equation

$$P = A + B \, \mathrm{Log}_e \, (\mathrm{Dose})$$

$P$ = Probit
$A$ and $B$ are constants

**Table 16.3** Some typical hazardous doses based on probit values.

| Cause | % Fatality 1% | 50% | Variable |
|---|---|---|---|
| Thermal radiation | 1000 | 3000 | $kW^{4/3}$ s |
| Chlorine | 120,000 | 480,000 | $ppm^{2.75}$ min |
| Ammonia | $2.6 \times 10^{10}$ | $2 \times 10^{15}$ | $ppm^{2.75}$ min |

$$\text{Dose} = \int C^n t$$

There are equations for thermal radiation, explosions, chlorine, ammonia, phosgene, etc.

To give some idea of the **likely** range of effects the following are some calculated dose values from probits. These are indicative not absolute values. These are not quite the same as given earlier as they are derived in a different manner based on the enhancement of the true dose determined by the area under the curve and not the TWA (Table 16.3).

Thermal radiation is also a 'dose effect' that is the effect is a function of time in this case the effect is

$$\text{Flux}^{4/3} \times \text{time}$$

Tables are available for pain and skin damage. As a person is retreating the flux will be reducing but the dose is still additive. Heavy clothing (as used by Firemen) is beneficial.

### *Design it out*

This is inherency and the following may give some clues.

### *Noise*

Noise can be reduced by design. It may not be perfect but the design of gears and turbine/compressor internals is an example.

Noise from high pressure let down, particularly gases, can be reduced by fitting control valve internals.

Once again these may not be perfect and other palliative methods may be required.

### *Leakage*

Leakage can be reduced by the correct choice of pump and compressor seals.

All welded piping is inherently safe.

Attention to the sealing of flanges and the gasket materials.

The selection of instrument fittings.

Once again these may not be perfect and other palliative methods may be required.

### *Reduce the impact*

#### Noise

Acoustic booths can reduce noise. (This may create its own problems for entry to the booth.)

Heavy-duty lagging (a form of damper) can reduce noise.

**Leakage**

Consider forced (or induced draft) ventilation in booths or over open tanks venting to a 'safe area'.

Review secondary seal/capture at flanges.

Mechanical sealing is inherently more leak proof than packed seals.

## 16.2 Human factors/reliability

It is arguable that the human is potentially a weak link in any identification or assessment of 'hazards', be it in a form of LOPA, Bow Tie or Fault and Event trees. It is essential that this form of vulnerability is analysed with care and detail.

HSG 48 Reducing Error and Influencing Behaviour has identified three main causes of human failure in its figure 2.

### Skill-based errors

Many of the activities we carry out are extremely complex yet are performed with little or no conscious effort, e.g. walking, driving, typing. Such tasks are very difficult when first attempted but are developed by practice into operations, which can be done repeatedly with a very high success rate. A mental routine is developed and stored and then recalled when required. Hence we can think whilst we type and talk whilst we drive, usually without impairing the basic task. This mode of skill based action is common to most repetitive tasks and hence many operations on industrial processes come into this class. Failures during skill-based actions are usually described as slips or lapses. They are unintentional and the frequency is modified by local conditions and the immediate circumstances. Typical slips and lapses include: failure to make an intended variation from the usual pattern. The regular pattern is followed past the point of intended variation. Omission of a step in a sequence, often associated with an interruption. The sequence is picked up at a later stage. Mixing elements of two processes, which are being carried out simultaneously. Repetition of a step. This may occur when there are waiting periods in a sequence and a check is made to identify where to resume. An error here may result in resumption too early in the sequence (repetition) or too late (omission).

### Rule-based mistakes

The next level of operation—rule based—occurs in situations, which are not completely standard and so are not fully covered by frequently used operating patterns. Instead, conscious analysis and a decision is required. It has been found that the mind analyses the information available and seeks a familiar pattern. If one is found that fits many features of the situation then the corresponding operating pattern is applied. Characteristic errors in rule-based operations include: misapplication of a 'good' rule. There is a tendency to try out a rule, which has proved to be generally good in other circumstances—particularly if it is a frequently used rule—even though it doesn't explain all of the information. Furthermore, individuals find it difficult to drop their hypothesis until there is overwhelming evidence against it—'mind set' occurs. If a signal indicating no flow has always meant that a control valve is closed it is hard to consider the alternative that the signal is faulty and the control valve is still open. Use of a wrong rule or a poor rule. A wrong rule may be chosen due to ignorance or misunderstanding of the system; a poor rule may develop from experience, as with ship's officers who permit encounters below the recommended separation distance since they have found from experience that they can normally manoeuvers to avoid collision. Thus higher-risk situations are allowed instead of being avoided.

## Knowledge-based mistakes

The highest level of operation is knowledge based where the individual finds no convenient previous pattern will fit and has to analyse the available information to decide on an actions. This is demanding and situations may easily become so complex that success is unlikely and a mistake is made. An important factor influencing the success rate is the operator's model of the process and knowledge of its current state. This is obviously greatly influenced by the normal interactions between operators and the system, as well as the nature of the information provided, as well as the manner of presentation. Problems with knowledge-based working include: limited mental processing capacity resulting in selectivity in the use of information; bias in favour of easily recalled data; resistance to changing a plan formed on the basis of an early hypothesis. In addition to the above error types—labelled as slips and lapses or mistakes—there can be violations, where a recommended procedure is deliberately not followed. Violations may be done with good intent, in the belief that better results will be achieved with a varied procedure, or when an unusual situation exists for which there is no obvious procedure. Violations may also occur for less honourable reasons, simply for convenience or to avoid discovery of a mistake. These are greatly influenced by the degree of supervision and by management attitudes - both implicit and explicit. If it is known that management will turn a 'blind eye' to violations in order to save production or to speed up maintenance, then violations are much more likely.

There is a general belief that some people are 'accident-prone'. This can be debated long and hard, but it is more constructive to analyse some of the possible causes. These can be broken down into many categories; the most common three are:

Experience and skills
Psychological make up
Stresses

## Experience and skills

The training and knowledge of the person must reflect the job description but also that person must have certain extra skills, which will reflect the need and ability to make decisions.

A good Engineer should have a good grasp of his/her own discipline and also a good understanding of the needs of other Engineering disciplines. There is a balance between depth and breadth of knowledge—the knowledge of the individuals' discipline has to be deep but there has to be breadth so as to work with other disciplines and understand/

interpret those other problems and needs. The good Engineer has to have deductive skills to recognise where problem areas may arise and the technical skills to overcome them. If either is missing there is the risk of an error/mistake. This not only applies to the technical problems but also dealings/supervision/checking of members in the work group. In the case of Management of Hazardous Plant, the Engineer must understand how the process works and also understand the weak points in the process. These skills take time and training and require the ability to use deductive skills, which convert technical knowledge into a practical application. (Most of this is 'knowledge based'.)

Operating personnel require similar training, knowledge and skills are still required but at a lower level. There may be situations that the operating personnel have not met or cannot analyse the event. If the operating procedures do not give clear guidance to the operating personnel, the personnel may make mistakes (panic), as they will be forced to operate outside the envelope of their experience and skills. This reflects back to the concept of *culture* and forward to *Managers Responsibilities*. (This is skills based.)

Don't blame an operator for making an error of identification if the order is A/B; A/B; A/C/B, there are 'expectations', which will become 'rules' in the mind of the operator. If an operator has to reach out over a handrail to access a valve it is human nature to ease the damage potential to the back by climbing over the handrail. The perceived risk of falling may be less than that of injury to the back and a precedent is set.

Human ergonomics, as a means to reducing 'mistake potential', involves layout but also size/colour, postural position, eye position and many other important features.

There are a number of references to human reaction under pressure/panic conditions. Lars Weisaeth has written extensively on this topic, one worthy of more study is 'Technological Disasters Psychological and Psychiatric Aspects' 7th International Symposium on Loss Prevention and Safety Promotion in the Process Industries Taormina, 4–8 May 1992. In this, he discusses perception of response after being put under stress or Post Traumatic Stress Syndrome/Disorder—and 'Flight from an emergency' (fight or flight); under these circumstances survival is more imperative than the logical correction of the fault.

D.E. Embrey discusses 'Human Errors' in a paper at the 7th International Symposium on Loss Prevention and Safety Promotion in the Process Industries. The best prediction is that all humans are 'error prone' and the objective is to reduce the scope or the consequences of this error (a flaw in the psychological make-up).

It is now recognised that what was called 'shell shock' (now recognised as Post Traumatic Stress Disorder) has some variations one being that of guilt. Guilt that 'your mate was killed but you were not'. Guilt that you did not do enough to help your mate. 'If I had tried to save him he would be alive now'. (This ignores the logical analysis of the reality of the situation that both would have lost their lives!)

British Midland Airways **Boeing 737**-400, G-OBME at Kegworth on 8/1/89 B737 was caused essentially by a reconfiguration of the pilot's displays. The pilot assumed that the configuration was as he had always found it. This was a hazard that could have been averted by more familiarisation training. The root causes were a mix of human error and of course a jet engine failure. There are a number of sources of information on the web.

## Psychological make-up

There are two forms of psychological make-up worthy of note. The first is the person who cannot handle stress of any sort and goes into a panic at the first hint of pressure. This person is likely to be error prone. The second is the one who will not listen to guidance and will do what he/she wants to. This is likely to lead to errors of judgement.

## Stresses

Stresses can be various and could include

Fatigue
Emotional
Work related

The first two are self-evident; as the human becomes more tired the judgement becomes flawed and mistakes result as the deductive skills become flawed. Emotional stresses could arise from problems at home or personal who are set upon by a work mate. Feelings of oppression and inferiority can lead to judgement being flawed. Good managers should be able to detect this through day-to-day workplace discussions/talking during 'walk about tours'.

Work stresses are also understandable. The human performs best under a slight or low level of stress. High levels of stress can impair the judgement and deductive skills and low levels of stress can produce a lethargy, which ignores information and in turn leads to lack of judgement or response. But there are more insidious stresses caused by poor ergonomics. Aircraft designers have worked hard to develop the Head up Display (H.U.D.), which is also becoming used in cars. Without realising it, the personnel can be put under stresses by the layout of the pages on Distributed Control System Screen, the position, order; ease of recognition/reading is essential. (See also Texaco Refinery Explosion and the Kegworth B737 crash.)

Other issues to be found in the brain are summarised as they are not easy to explain in simple terms.

## Information overload

In Information overload the brain has TOO MUCH information and cannot sift the critical or top-level information from the low-level unimportant information. In effect the reasoning powers are swamped by both essential and trivial information and so the outcome is that nothing is done. This can be analogous to a juggler—there is an absolute limit to the ability to handle the number of objects and beyond that limit things get dropped.

The concept of information overload can be dealt with by two strategies. At one level the operators have sufficient resources to handle all of the work and at the other, the information is filtered and presented on a clear unambiguous form (skills and rules). In process plant it is not only the information but the size; on a small plant where the transit time may be small the supervisor may be able to handle more as there is less time used in moving from A to B to C in data collection. Above all the presentation of clear unambiguous data with the appropriate diagnostics in a Control Room is fundamentally important. The human can only accept a limited amount of information at any one moment and the message must be clear and unambiguous.

Training and background knowledge all help to reduce the potential for information overload as also does practices. There are no solutions or fixes; an understanding is required as well as the open mind and eye. The key question must be:

'How could I handle the problem/problems professionally and without error?'

There are a number of informed studies on this subject, which can be summarised below.

Is the MOST important information displayed as a priority?
Is it clear and unambiguous?
Are there clear instructions on what to do if that sequence arises?
Has training been given on that sequence?
Has the data flow been subjected to a rigorous audit?

## Mind set

The person has a fixed idea and cannot be convinced that there may be an alternative explanation or idea. It could also be called 'tunnel vision'.

There are a number of informed studies on this subject, which can be summarised below.

Has the technical competence of the individual been tested?
Has the person been adequately trained and does that person have demonstration of competence?
Is there some underlying psychological issue?

## Cognitive dissonance

This is quite difficult to explain. The mind tries to fix the evidence into a picture. Some does not fit so is rejected or reasoned away. The brain is quite convinced that the evidence is now consistent but ignores the fact that some key evidence may have been rejected or distorted due to some erroneous logic.

Has the technical competence of the individual been tested?
Has the person been adequately trained to analyse issues correctly and has that person demonstrated competence?
Is there some underlying psychological issue which might preclude that person from a role?

## Panic

The person just cannot make any decisions.

Has the technical competence of the individual been tested?
Has the person been adequately trained to analyse issues correctly and has that person demonstrated competence?
Is there some underlying psychological issue which might preclude that person from a role?

## Confusion

An instruction that is not totally understandable and in simple language has the potential for confusion or misinterpretation and may lead to errors as will panic.

It may be that confusion is the result of poorly written instructions, which leave a person to deduce what is required (see App 5 HAZOP Guide to Best Practice Ed 3). If this is a common feature in an organisation due to culture or poor supervision this may result in a 'common cause effect', which is different from a 'common mode effect', which is equipment related.

The military practice issues such that they can react correctly without deviation in an incident. There is more a reflex action than a determined action.

## Complacency

This is best summarised as '**It cannot happen here as it has not been seen here**'. This feeds back to the need for Audits of procedures and safety management systems (Chapter 17).

## Training

While there are a number of contributions to 'Human Error' as shown above, it is equally important to carry out training on handling emergencies or the unknowns and to carry out practices. This is clearly illustrated by both the military and the emergency services.

## Other readings on human factors

Health and Safety Executive, 'The explosion and fires at the Texaco Refinery, Milford Haven, 24 July 1994: A report of the investigation by the Health and Safety Executive into the explosion and fires on the Pembroke Cracking Company Plant at the Texaco Refinery, Milford Haven on 24 July 1994', ISBN 0-7176-1413 1, 1997.

While the previous sections have addresses the more technical aspects of the vulnerability of humans no one should ignore the impacts of noise (tinnitus), vibration (vibration white finger) and thermal radiation not to mention also the extremes of heat and cold.

## 16.3 Human factors—An example

This is an attempt to illustrate just one of human factors, mind set, during a new process plant start-up with novel features illustrated to show how errors can arise from lack of experience (knowledge).

The supervisors and operators were picked from an experienced group of personnel, so the experience will be satisfied by years of training. However, the new plant had both unknown and un-experienced characteristics. The team were alert to their lack of experiences drawing on their previous experience and detective skills.

### The process

Acetylene hydrogenation reactors require passing a mixture of hydrogen, ethane, ethylene and acetylene across a fixed catalyst at about 60°C in a single pass. The reaction is of course exothermic and the catalysts are both reactive and selective, that is, it has to be raised to an initial temperature (60°C) before the hydrogen/acetylene reaction takes place. There are two potential catalysts, one is used at the front end of the process where there is excess hydrogen in the process fluid and uses the thermal mass to maintain a low-temperature spread so avoiding the ethylene hydrogenation phase. The second catalyst, used at the back end of the process, is more efficient and uses a fixed ratio of hydrogen to ethylene but beyond a threshold temperature the hydrogen/ethylene reaction also takes place and is favoured. The Start-up Manager had experience of the first, simple mixed-stream Hydrogenation process, but not the back end process. The Start-up Manager raised the inlet temperature rapidly with an excess of hydrogen such that the reaction took place in a limited section at the inlet to the bed. This produced a rapid rise in temperature, which produced reaction conditions selective to the hydrogenation of ethylene and the acetylene concentrations in the reactor exit were out of specification. The immediate assumption was that the catalyst was faulty. A second manager was more careful and ensured that the reaction spread evenly throughout the reactor bed such that the total acetylene conversion was achieved within a few hours.

The first manager had entered a mind set.

## 16.4 Vulnerability equipment

### Effects on materials

The effects on equipment should be covered in Hazard Studies (Chapter 3), HAZOP (Chapter 4) and HAZID (Chapter 5) and also where appropriate in FME(C) A (Chapter 12). However it is necessary to understand some of the equipment vulnerabilities.

In addition it is important that an experienced metallurgist is involved in the selection of materials of construction.

Materials are subject to assault just as are humans. The following is a guide to some of the more common forms of failure.

Steel loses most of its strength above 450°C in fire assault. Some possible but not all-embracing outcomes are:

- Once a structure softens and collapses there is the potential for a domino effect.
- Once a vessel, or system under pressure, is challenged there is the potential for a boiling liquid expanding vapour explosion (BLEVE) or rupture if it is not de-pressured or protected against fire by deluge or lagging.

Machinery and structures are displaced by blast waves. This can be a source of escalation from broken joints or sealing systems.

Blast walls are a twin-edged sword (beware of poor dispersion created by the enclosure).

Steels at elevated process temperature are subject to creep. Exotic steels may be required and the temperature limits set. The creep life may be seriously reduced by brief excursions beyond these limits.

Some steels are subject to brittle failure at low temperatures. Special steels may be required and attention paid to start up and shut down conditions during depressuring.

Steel is subject to internal corrosion due to:

crevice corrosion, galvanic action between dissimilar metals and corrosion under debris (a variant of crevice corrosion)
external corrosion (corrosion under insulation CUI from rain water or deluge water).

Steels are subject to erosion with slurries and trace solids in the flow regime.

Some steels are subject to stress corrosion cracking (SSC)

chlorides in the case of stainless steels
nitrates/nitrites in the case of carbon steels (see Flixborough).

Steels are subject to 'metal fatigue' following many reverse loading cycles. This may be caused by:

vibration or pressure cycles or overhung (seismic) loads and flow or reciprocating compressor induced vibration.

Cylindrical bodies are subject to flow induced vortex shedding. Examples are:

poorly designed heat exchangers are subject to fatigue due to vortex shedding from the tubes at different parts of the flow regime
thermo-pockets are liable to vortex shedding
stacks and chimneys are subject to vortex shedding.

Long pipelines are subject to 'water hammer' following sudden closure of an exit valve.

Steam mains are subject to 'steam hammer' if not warmed up properly.

Debris may be deposited in vessels, distillation columns and instrument tappings.

Debris may choke control valves and also build up on internals so making them difficult to operate.

Water may freeze and waxy fluids solidify in winter. Note also instrument tappings.

Bearings are one of the weak links in any system. Lubrication is essential but consider also:

rolling contact bearings may fail in fatigue mode, 'false brinnelling' due to poor fitting techniques and interference failures due to 'wear' on shafts and castings

journal (white metal) bearings may fail in fatigue (under specific vibration conditions) or wear and tear due to poor filtration of oils.

Pump seals may fail due to:

poor fitting and assembly, debris in the process fluid and wear and tear. (The seal arrangement is a special study in itself ).

Compressor seals may fail due to many causes and are influenced by the design features and arrangement of which there are many. (The seal arrangement is a special study in itself.)

# 17

# Safety audits

## Chapter outline



## 17.1 Definition

An audit is an examination of the systems of work of an organisation. The Health and Safety Executive describe [1] safety audits as: 'The collection of independent information on the efficiency, effectiveness and reliability of the total health and safety management system and drawing up plans for corrective action'.

## 17.2 Description

Auditing is a common and widespread method [2–4] used in many varied, major industries, especially in finance. A safety audit is a structured process whereby information is collected relating to the efficiency, effectiveness, and reliability of a company's total health and safety management system. They are routinely conducted in order to determine whether the company is in compliance with safety legislation. An audit has a number of features:

- It is systematic
- It has a clearly defined objective

A Guide to Hazard Identification Methods. https://doi.org/10.1016/B978-0-12-819543-7.00017-3

- It must have a scope—the focus of the audit—and boundaries
- It must have terms of reference
- It is based upon sampling key parameters/features and does not examine every feature within the scope
- The recommendations must be measured against certain parameters or reference points such as 'best practice'

## 17.3 Resources

*Manpower*: The depth and nature of the audit will influence the manpower required. A typical number is 6–8 and it can be an advantage to have an even number of members. The list of team members has to be finalised and agreed with the initiating manager with the team chosen according to their skills and the nature of the audit itself.

*Planning*: The detail, which goes into the planning and preparation, can pay dividends during the audit, not least on the 'domestic arrangements', if costly delays are to be avoided. These will include:

- Accommodation—hotels, etc.
- Secure office
- Telephone
- Filing cabinets for storing documents
- Secretarial support, including photocopying
- Identity badges while on the site
- Liaison person on the site

## 17.4 Timing

Audits should be a regular and routine part of any project life.

## 17.5 Advantages, disadvantage and uncertainties

### Advantages

- There will be a steady improvement in corporate culture leading to lower incidents/ accidents.
- They provide a reminder of the importance of regulations and safety standards.
- They can highlight trends—good or bad—throughout the lifetime of a plant.
- Any audit can be made as flexible as is required.

### Disadvantages

- The audit might create frictions between members who are being audited.
- The site personnel may not accept the recommendations.

### Uncertainties

There is the inevitable risk that a team member may have some fixed ideas about what is 'best practice' particularly if the member is not familiar with that system. This should be handled at the team selection stage (Section 17.3).

## 17.6 Applications

Audits can be used in many ways to improve the safety (and environmental) performance of a company. They may be at one level the audit of the Safety (Environmental) Management Systems and at another the monitoring of workshop procedures. One obvious cause for an audit would be the manner in which Safety Systems were tested and both 'how' and the 'frequency'. The potential applications have no real boundaries.

### Background

Audits can be used for the identification of weaknesses in both safety and environmental systems, which may result in a hazard to employees or the community. Whilst not a prime method of hazard identification an audit may find additional problems that were initially overlooked, have crept in due to changes in a process or in the operating methodology, or have re-emerged due to familiarity, ignorance, short cuts and neglect of good practice or poor maintenance.

*Types of audit*

Audit teams must be independent in order to avoid conflicts of interests or, even worse, prejudices. The team may be external or internal and, in the latter case, either from on-site personnel or from another site or department within the company.

- *External.* Here the auditors come from outside the organisation. There are obvious advantages since this ensures total independence and provides a different perspective on issues. The auditors may also be highly skilled and experienced in the technique, knowing what and where to search for 'tell tales'—indicators of problem areas. Also they may have more time for in-depth examinations than those who work on the plant.
- *Internal—local.* In this instance the auditors will come from the local organisation. This grouping will probably not have total independence and may have prejudices, positive or negative, and the team may lack the skills required to see the problem areas.
- *Internal—corporate.* Here the auditors come from another site or department within the corporate organisation. This grouping provides greater independence and, probably, fewer prejudices and more skills although an element of 'corporate blindness' may remain. A positive factor can be the transfer of good practice from one site to another.

The type of team that is chosen should reflect the terms of reference and scope of the audit. Thus whilst it may be appropriate to use an internal audit team to examine the compliance of a particular plant with the Permit to Work procedure it would be advisable to use an external audit team to examine the corporate management structure.

### *Audit steps*

There are a number of clearly defined steps in an audit:

- Setting the terms of reference and scope
- Planning and preparation
- Preaudit sampling/data gathering
- Organisation based sampling/data gathering
- Forming a judgement
- Drafting the report
- Issuing the final report
- Follow up to the report

The audit must have some key objectives such as:

- Identification of the strengths (and weaknesses) of the local or corporate Safety/Environmental Management Systems, (S/E.M.S.)
- Identification of improvements considered necessary to improve the S/E.M.S.
- Establishing an action plan for the improvement of S/E.M.S.

Those must all be measured against a set of references, these will include:

- Corporate policy
- Corporate standards
- Legislation present and future
- External standards
- Reasonable best practice

### *Sampling data*

The introduction indicated that the audit is a sampling exercise. It is not possible to check every item in a batch and it is necessary to select a sample for detailed analysis. The sample size must reflect both the size of the population and give confidence that the sample is truly representative of the whole. If the total population is 2 and the sample is only 1 it would be wrong to conclude that if the one sample of item was 'healthy' the other was also 'healthy', but if the population was 1000 a smaller percentage of the sample would give good indicators of 'health'. Various sampling factors have been suggested. The following are indicative values.

| | Sample size for confidence levels | | |
|---|---|---|---|
| **Population** | **98%** | **90%** | **85%** |
| 10 | All | 9 | 8 |
| 50 | 45 | 30 | 16 |
| 100 | 80 | 40 | 20 |
| 1000 | 300 | 60 | 20 |
| 2000 | 320 | 60 | 30 |

Taken from IChemE Training package Safety Audits Courtesy IChemE.

It is clear that a larger sample is needed if the auditors wish to have a higher confidence of their conclusions. However time and resources will affect the choice of the sample size.

### *Identification of a 'noncompliance'*

This is central to the audit process and also requires the most skill. This process can be described as three questions:

What?
Where?
Why?

The WHAT question requires an analysis of both expectations and required standards. The expectations are that good working practices and management systems are in place in the particular activity. The required standards will flow from corporate policies and the legal/regulatory requirements. Some expectations will flow from these requirements and some from experience. This is a very open statement and requires illustration. The Health and Safety at Work, etc. Act requires 'the provision and maintenance of plant and systems of work that are, so far as is reasonably practicable, safe and without risks to health'. It does not call explicitly for a Permit to Work but on a process plant they will be expected as one of the 'systems of work'. In contrast, in an office a permit to work would not necessarily be expected. On a construction site, particularly a green field site, tractors might be expected but the access of tractors to an operational oil refinery would not be expected. The identification process requires a large measure of experience, knowing WHAT to look for and WHERE to look for it. Sometimes check lists may usefully help, either standard lists or ones prepared for that specific audit.

When walking around a process plant there should be a mental (not written) checklist. If one is used as a rigid crib sheet the auditor may be acting mechanically and not taking a questioning approach. The table below gives some examples of how a good auditor might observe and think.

| | **Task** | **Reason** |
|---|---|---|
| a | Check safety gates open/closed/stiffness | Do plant personnel visit the area of the plant? |
| b | Are relief valves properly supported against jet reaction? | Relief valves branches may be bent by jet reaction forces and if it is bent the relief capacity may be restricted. |
| c | Are static waste skips on the site? | Do vehicles have to come into a hazardous area to recover the skip? Is it a source of fire/flame? What controls are applied for their safe removal? |
| d | Are steam relief valves tail pipe drains clear? | If the relief valve lifts someone may be burnt with hot condensate. The capacity of the valve may be compromised. |
| e | Are firewater drains clear of debris? | If they are fouled there is a possible risk of the escalation of a fire due to fire spread upon the un-drained water. |

*Continued*

| Task | | Reason |
|---|---|---|
| f | Are all sources of liquid spill or leakage handled effectively and fully contained? | Leaks may enter the water course and hydrocarbons may accumulate in the soil to be floated out by the application of fire water at a future date. |
| g | Is cellulosic/waste collected in closed containers? | Waste may blow out of the site into the public area. |

The above list is only indicative of the items in a possible check list for a site tour. In the next table there are a few examples of other clues, which might suggest further investigation is needed.

| Observation | | Possible implication |
|---|---|---|
| a | Damage to curbing inside hazardous areas with tyre marks nearby | Is there unauthorised traffic in the hazardous areas (a source of ignition)? |
| b | Tour guide 'appears' to be 'leading' you from an area of the site | What is being 'hidden' and what should you not see? |
| c | Damage to lagging or rust at the bottom of the tank | Possible corrosion under insulation leading to leaks |
| d | Roped off areas with no signs of maintenance | Is maintenance being underfunded and is there a latent problem? |
| e | Oil stains on roads | Oil is entering the drains and bypassing interceptors |
| f | Open sample points | Vapour release to atmosphere |

The WHERE question is seeking to identify the source of the noncompliance, or the deviation from expectation and required standards within the organisation or its management systems. The problem may lie in:

- The organisation
- The personnel
- Other

The WHY question aims to seek out the root cause of any noncompliance or deviation from expectation or required standards. These are likely to be fairly clear and have fairly clear indications; again the list is obviously long and again only illustrations can be given.

The following are some indications of the 'what' and 'where' with respect to the organisation, which can arise during an audit.

| Symptom | Organisational problems Possible cause | Consequences |
|---|---|---|
| People feeling uncertain and isolated | Lack of personal objectives | Personnel do not know what to do in specific situations |
| People feeling they are given poor guidance | Lack of supervision or poor procedures | Bad decisions are made leading to potential hazards |
| People feeling that any message is conditional, 'of course safety is important **but** production comes first'. | Perceived lack of management commitment | Bad decisions are made leading to potential hazards |
| People have poor skills | Lack of training | Personnel make unnecessary errors |
| People feel uncertain as to who is their Supervisor/Manager | Poor reporting routes, personnel problems | Bad decisions are made leading to potential hazards |
| New recruits do not know what to do | Lack of training | Bad decisions are made leading to potential hazards |
| Head strong personnel doing what they think is right—not what they are told to do | Corner cutting/ lack of discipline or control | Bad decisions are made leading to potential hazards |
| Personnel feel they are being victimised or personnel having no respect for the manager and ignoring guidance | Personality clash | Bad decisions are made leading to potential hazards. An alternative consequence is that the person is incapable of making a decision |

Other problems could be poor accident investigation or inadequate follow-up from the investigation of an accident or a near miss.

### *Formulating a judgement or conclusion*

This is the final task of the audit team. The judgement or conclusion will lead to a finding and then to a recommendation and it is therefore essential that the recommendation is appropriate to the issue.

First the judgement must reflect the needs and product of that company as well as its corporate structure. This can best be described by simple examples. A research company, which produces a technical product does not necessarily require a Permit to Work system, but, if it is of sufficient size it may be appropriate to expect one person who has a responsibility for safety, be this good housekeeping or for the practice of office evacuations. A workshop, which employs many mechanics, may justifiably have one person responsible for the Occupational Health and Safety of employees. The judgement must therefore reflect the expectations of the Auditor and be in harmony with the structure and type of organisation being audited.

Second the judgement must be measured against a reference. This reference may be a corporate standard or a legal requirement: these should be clear-cut.

Third the judgement must be fully researched and supported. One person who did not wear head protection or one spillage would not be a suitable sample and may be due to one rogue employee—it is appropriate that this is highlighted but it is not a major finding. However, if these examples are prevalent it may be necessary to research it further to find the real source of the problem.

## The audit process

The main steps in the audit process and the essential tools and skills have been described. In this section the full sequence is considered in its logical order.

### *Terms of reference and scope*

Normally a manager within an organisation (design, operations research, etc.) will ask for a Safety/Environmental Audit for one of a number of reasons. It may be that an audit is required on a fixed routine as specified in the Corporate S&E Management systems or a manager may feel that there is a short fall in the safety/environmental performance and an audit would help to identify the problem area. It is important that it is requested and not imposed. The manager will call for the audit to be carried out and a leader or chairperson will be nominated. The manager and leader will then decide upon and write the terms of reference and the scope of the audit. Next they will agree the type of audit to be carried out and draw up a possible short list of team members.

The terms of reference will generally cover:

- The type of audit to be carried out
- The objectives of the audit
- The start date of the audit
- The delivery date of the audit
- The person to whom the audit should be delivered (if different from the initiator)

The scope will generally cover:

- The areas included
- The areas excluded
- Any specific area for detailed attention
- The references against which the audit should be measured

Preparation for a major audit may take over a week but for a local audit it is only hours.

### *Implementation*

The team should meet to make their acquaintance and that of the liaison person if possible, as they may not have met before. Following from this a list of initial documentation should be drawn up for review. The documents should be allocated amongst the team. The team should consider which items need special attention and draw up an initial list of:

- Issues to be examined
- Personnel to be interviewed
- Any check lists they feel appropriate for the future audits

They should then produce an initial timetable of tasks and interviews.

Preaudit sampling/data gathering: Following the initial meeting the team may need a few days in which to review documents in their own office and analyse the data already delivered prior to the site visit. It is very possible that extra documents or data will be required following this initial review. This should be collected and forwarded via the liaison person, followed if necessary by a further meeting of the team.

### *Organisation-based sampling/data gathering*

The first on-site meeting of the team must be a 'courtesy meeting' with site personnel. During this meeting the team and site personnel should be introduced to each other and the terms of reference and scope explained. Whilst it would be usual for the site personnel to already know the terms of reference and scope it should still be possible to expand on them and to include any particular topics which the site personnel may feel appropriate.

Following the meeting Identification Badges must be issued and worn and the team should then be given a site safety induction briefing and a familiarisation tour round the site to examine the detail and potential hazards of the operation and also to be introduced to other site personnel. This tour may need to be supplemented by other, more specific, tours at a later stage.

In the initial phase of the audit the team are basically searching the organisation using the **WHAT**? **WHERE**? **WHY**? approach. The method may be by interview, examining records, data collection and analysis or by the well-trained eye and ear. There are no hard and fast rules as much is based on experience—it is surprising how often a finding is identified by an experienced auditor.

At the end of each day a short team meeting should be held to exchange observations and classify details. The meeting need not be highly structured but would generally include:

- Provisional, preliminary findings (these may change at a later date)
- Areas requiring further clarification
- Gaps in knowledge
- Areas needing further study

At the beginning of the day and not longer than every other day the leader should meet with the liaison person and a small group of the site personnel to give them feedback of the provisional findings. During these meetings any misunderstandings can be cleared up and the site personnel will be prepared for the final report. Initially progress may appear to be slow and it may require further visits to the site to assist in the data gathering and formulating the findings—the audit team should not be put off by what might appear initial slow progress as it is not unusual with the team on a learning curve. It is also very important that all data are logged and recorded for reference purposes and filed in a suitable manner. The data become the evidence, which supports the finding and recommendations, and unless it is available, confirmed and secure then the findings and recommendations cannot stand.

### *Draft report*

If possible the draft report should be available and presented to the site personnel before the team departs. Inevitably there will be negative features in the report, so it is important that these are counter balanced by the many positive aspects that were found. If the feedback process has been successful there should be no surprises in the report. But if the feedback process during the audit process has not worked there is the risk of undermining the whole validity of the audit with the killer statement 'did you not understand/fully investigate the issue - surely you know that…'.

### *Final report*

The final report must reflect the observations of those being audited. If the appropriate liaison has been in place and supporting information has been available then the final report should be only a polished version of the draft.

### *Follow-up*

The terms of reference may require a follow-up to check that the recommended actions have been effectively implemented.

## Techniques for gathering data

An important feature of audit meetings and investigation is the approach of team leader and the members. Whilst it is not appropriate to cover it here in detail it is important to mention the subject. Auditors are eliciting information from the organisation and the personnel, and to a degree the interviewee will be defensive, so interpersonal skills are important. Auditors must be courteous and nonaggressive. Questions should be open, enquiring and wide-ranging, not just answerable by a yes or a no. Interviewees must be allowed adequate time to talk and respond and unsatisfactory or incomplete answers must be further probed. Auditors need to be good listeners, open-minded and nonjudgemental. The auditor should have only 25% of the talking time and the interviewee 75% of the time.

In many cases the eliciting process will probably involve asking two questions. The first a benign question which allows a more directed question if the reply gives cause for concern. An aggressive first question will achieve nothing.

In order to avoid conflict—as there is a potential for conflict if the interview is not carried out with sensitivity—the auditor **must**:

- at all times, act in a professional manner,
- whenever possible, be aware of the potential for conflict and try to avoid it at all times. He/she should, where possible, be careful of the dangers of putting someone in a '***lose/ lose***' position,
- be a good listener and able to analyse what is being said quickly,
- be able to formulate questions in an easy to understand, unambiguous manner,
- be open-minded and nonjudgemental,
- be open eyed and
- be open eared.

## 17.7 Examples

### A simple audit

A simple audit that is easily carried out is of the execution of the Permit to Work (PTW) system. A Plant Manager (Production Manager)—who might be both the Initiating manager and auditor—could carry this out.

The terms of reference could be

- To examine the PTW system to ensure that it is carried out according to works/site procedure.
- To deliver the results to the plant supervisor (name) by (date).

The scope could be

- To examine a sample of PTW for the weeks x/y.
- To include hot work and cold work permits but to exclude entry permits.

The auditor must now decide on the sample size—possibly 50% for a total of 100 permits and then examine the permits covering:

- Were they filled in correctly and unambiguously?
- Was the task or job clearly specified?
- Were the isolation standards clearly specified?
- Were the precautions (risk assessment and site preparation) clearly specified?
- Were the personal protective systems clearly specified?
- Were the tools clearly specified?
- Were rules set out when or why the work must stop and the site made safe?

In an ideal world it would be useful to observe the operation at a distance to ensure that there was compliance with the requirements of the permit. This may create frictions, which would be difficult to erase.

## A more complex audit

This may be recommended, for example, if there are reasons to believe that, as a result of a number of incidents, there is an organic problem, which might be uncovered by a focussed audit. Such an audit may lead to the identification of the source of potential hazards, which is the title of this book.

### *Terms of reference*

To examine the accidents records for the years 2000 and 2018 to identify whether there has been a systematic drift and to identify improvements to:

- The management structure
- The supervision
- Training and experience standards
- Deliver the report to person A by date B
- Scope.
- To examine the accidents data for the years 2000 and 2018
- Use best standards as the reference

# References

[1] Quoted in ROSPA Health and Safety Audits e-book.

[2] HAZOP: Guide to Best Practice, third ed., Elsevier, 2015, p. 55 (Section 7.7 Auditing a HAZOP study).

[3] CCPS, Guidelines for Auditing Process Safety Management Systems, second ed., Wiley, 2011.

[4] S. Mannan (Ed.), Lees Loss Prevention in the Process Industries, fourth ed., Butterworth-Heinemann, 2012.

18

# Bow-tie diagrams

## Chapter outline



## 18.1 Definition

Bow-tie diagrams are a visual representation for a hazardous event of the sequence from the initiating causes to the range of consequences, including the worst-case outcomes.

## 18.2 Description

The central element of a bow-tie diagram is a hazardous event—the top event. Leading to this are the sequences from initiating causes (or threats) through the protection and prevention barriers that can stop that causing the top event. Extending from the top event is the full set of different consequences that could result from its occurrence. The fundamental intent of the bow-tie is to provide a readily understood picture of the full range of causes and the barriers in their paths leading up to the top event. When initially set down it is in narrative format, unlike fault and event trees, which have frequencies and PFDs assigned to each element. The expansion of the bow-tie allows a more critical assessment of the root causes and allows deeper analysis. Fault and event trees tend to use generic data/information, which may not be case specific.


**A Guide to Hazard Identification Methods. https://doi.org/10.1016/B978-0-12-819543-7.00018-5**

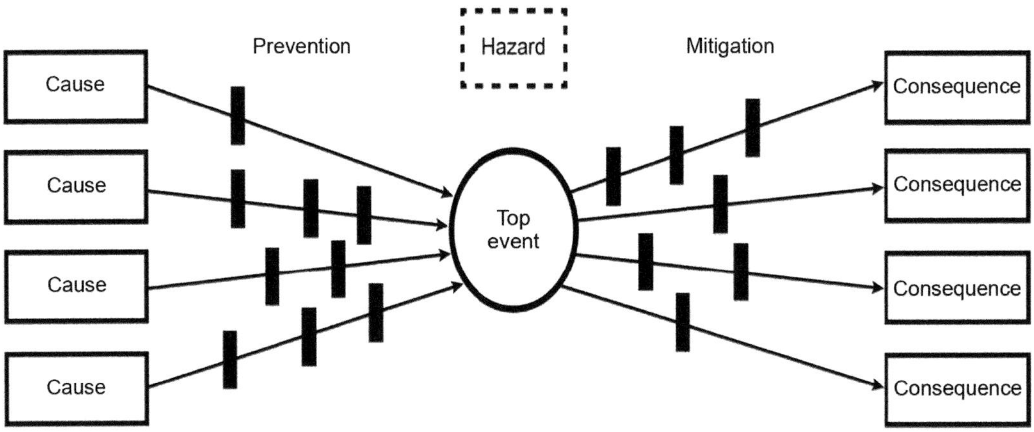


**FIG. 18.1** The generic form of a Bow-tie diagram.

The left-hand side of the bow-tie is effectively a fault tree rotated through 90 degrees, whilst the right-hand side is effectively an event tree. The basic pattern is shown in Fig. 18.1.

It is vital that the most serious risk is identified as the top event and that all possible consequences are included. Each path leading to a consequence shows the mitigation barriers that could, in this instance, reduce the effects. The topic is described in detail in the recent CCPS book [1]. The terminology and methodology of bow-tie diagrams has been covered by Manton et al. [2]. A useful approach is to ask 'What is the worst possible thing that can go wrong with this material or in this process?' When this is identified it is possible to search for what must go wrong in the process to allow it to happen. This makes is less likely that a major consequence is overlooked simply because it has never happened within the experience of the study team.

A particularly difficult challenge for the safety analyst is the full identification and assessment of rare events, especially those with major consequences. There is a tendency to remember and concentrate on events that have recently occurred and hence are at the forefront of the mind and of course these are seldom major in their consequences. There is often insufficient search of safety archives from related industries, which could give a more complete picture of all the possibilities. Methods such as HAZOP do not easily identify large-scale consequences, which come about through multiple failures. Even LOPA may not adequately identify all the major risks associated with a process.

There have been several instances that illustrate this problem, including the vapour cloud explosion at Buncefield [3] and the ammonium nitrate explosion at Toulouse [4]. The possibility of these major outcomes was not identified as the worst-case event and was not covered in the safety cases submitted to the regulatory authority. The first reaction when such a major event occurs is to label it as unpredictable (a 'black swan' event[a]) and

[a]Black swan events are those which are historically unknown and so are said to be unpredictable. The name originates in the 17th century belief that there are no black swans—ended by their discovery in 1697 in Australia (*Cygnus atratus*).

hence one that could not be anticipated. In the case of chemical accidents true black swans are rare—in both of the examples listed above a thorough search would have uncovered previous incidents. The question is whether the event is truly unknown or simply unknown to, or dismissed by, the analyst. This issue was publicised in by Donald Rumsfeld in his commentary on the evidence of weapons of mass destruction in Iraq in 2002.

| | *Knowledge: known* | *Knowledge: unknown* |
|---|---|---|
| **Awareness: known** | **Known**/*known* | **Known**/*unknown* |
| **Awareness: unknown** | **Unknown**/*known* | **Unknown**/*unknown* |

The known/known events are many and are likely to be included and eliminated in any thorough safety study, for example by design standards or checklists with backup from HAZOP study. Unknown/knowns events represent a failure of awareness of historical occurrences within or closely related to the process being analysed. Both the Buncefield and the Toulouse events were unknown/knowns and so could have been anticipated. It is vital that analysts recognise these categories, avoid having unknown/knowns. Known/unknowns should be rare but are capable of inclusion in analysis to minimise the possibility of occurrence or to be kept under review as knowledge and experience grows. The possibility of tipping points being reached in climate change studies is an example. With the wealth of experience from industrialisation there should be very few unknown/unknowns—an unanticipated major event that has never been experienced. Perhaps a What If study or use in HAZOP of the guideword What Else might prompt consideration. Any claim subsequent to an event that it was an unknown/unknown should be critically examined—it is a tempting excuse for an inadequate study.

In constructing a bow-tie diagram the first step is to identify the hazard and then the critical event—the top event. These may well be atypical events, unlikely to have been experienced recently, if ever, at the site or company. A historical search should be made as well as a realistic examination of the worst-case possibility for the process concerned. There may be more than one critical event to consider and a separate diagram must be created for each. The search must be thorough and honest to seek out any unknown/knowns and also to cover, as far as possible, the known/unknowns. At this stage the general details of the process will be known, there should be research into past events and discussion of various plausible consequences. This allows definition of each hazard and the related top event(s).

Now, taking each top event in turn, the separate possible causes must be identified and listed and then the prevention barriers identified for each one. Some barriers[b] may be common to several different causes. It is vital that a careful analysis is done to guarantee

[b]Most terms used in connection with bow-tie diagrams have already been defined and used in previous chapters. The term 'threats' may be used to represent a case specific description of cause. The term 'barriers' includes active and passive means of prevention development of a cause or mitigating the consequences.

that only truly effective barriers are written into each channel. On completion this gives a full fault tree for the top event. If it is quantified then the frequency of the top event is obtained by summing the frequencies for the individual channels.

Next the event tree must be constructed. There may be several different outcomes depending on the circumstance. For a flammable liquid this may be an unignited pool giving off toxic fumes. Immediate ignition could give a pool fire whilst delayed ignition might result in both a vapour cloud explosion and a fire. Domino effects might result in some cases. It is likely that there will be several routes leading to each of these outcomes. Every one of these must be examined to identify the mitigation barriers, if any, in each channel. The threats of injury or death(s) need to be estimated as well as the consequential damage and loss for each final consequence.

At this stage the basic bow-tie diagram of Fig. 18.1 has been developed into a usable description of the causes and consequence for the top event. There are some important checks that help to ensure a valid and useful diagram has been achieved. Each barrier should be examined to see that

- it can, on its own, prevent the progression to the top event or is able to fully mitigate the consequences or reduce the severity in a measurable way,
- it is fully independent and has no common failure mode with another barrier and
- it can be audited.

When quantification is done—and this is a necessary step in order that acceptability of the existing design can be tested against accepted criteria—it is essential that the frequencies of causes are not simply generic values but are appropriate for the plant under evaluation and also that the estimation of consequences is done for the actual plant and site. It is also important that the values assigned to PFDs of the protective barriers and for the effectiveness of the mitigation barriers are realistic, both as designed and as exists after a period of operation. Maintenance and testing of alarms and trips must be planned to uphold the specified effectiveness. Where human action is a part of the protective action consideration must be given to design of the equipment, the operating procedures and the selection and training of the operators in order to ensure that the predicted level of intervention is realistic [5]. Even the frequency of causes may change with time as plant ageing may affect mechanical integrity and hence loss of containment [6]. This requires that, as with any other hazard study, the bow-tie analysis is periodically revisited and critically reviewed.

## 18.3 Resource requirements

### Manpower

Whilst in principle an individual can carry out the analysis and construction of a bow-tie diagram it is better done by a small team to ensure that there is full appreciation of the hazards and the process details. As human factors may play a significant part in the sequences it is important that this can is properly represented.

### Data

For a process at the design stage the team should have access to all previous hazard studies and to the latest design details, including materials, amount, the intended operations and process conditions and the site and surrounds of the plant. Historical performance and incident records should also be available.

For an existing plant and process up-to-date drawings, operating procedures, maintenance and incident records should also be available.

### Software

It can be helpful to use commercial recording and construction software to assist the meeting recorder.

### Timing

A bow-tie analysis is usually carried out early in the project development. If the process integrity is based on a major piece of shutdown logic the most likely time is after the concept selection but before the detailed design. However it should be part of the regular review of the process and tasks. If the instruments are more standard or simplex it might be appropriate to delay the analysis until the detailed design is complete and has been reviewed by HAZOP study. It may be of use during the lead up to demolition.

Occasionally the need for analysis may arise from the HAZOP study itself or be used to support the concept selection where critical issues have to be dealt with.

## 18.4 Advantages, disadvantages and uncertainties

### Advantages

- A good bow-tie diagram provides:
  - a clear picture of the possible consequences and the routes by which they arise
  - the necessary conditions and sequence of events for each to occur
  - the relative importance of each safety barrier and the consequences of failure
  - the points where additional safety barriers are needed
  - the conditions requiring further in-depth analysis
- The analysis and its visual representation can help all concerned with the safety of a plant to recognise the sequences that could lead to a catastrophe and to appreciate the importance of maintaining the preventative barriers and mitigations that must be maintained.

- It is less likely that major consequences will be overlooked and fewer dismissed as 'black swan' events, i.e. those thought to never have happened and so cannot be considered.
- Regulatory authorities can be assured that a full analysis has been carried out and that the hazards are understood and satisfactorily controlled.

### Disadvantages

- The construction of a bow-tie diagram is potentially time consuming and should only be used in the appropriate circumstances. Each should refer to a specific problem.
- The analyst needs to be skilled in the use of bow-ties, particularly in determining the degree of detail to be included, and have a detailed understanding of the system under analysis.
- The value of a study will be limited if the available data are of poor quality and lack robustness or relevance. It is not a means of getting a precise prediction from imprecise data. Poor data may lead to an incorrect ranking of the risks.
- All results must be treated with care.
- The results should, for major events, be conditioned by consequence analysis and risk criteria.

### Uncertainties

- It is essential that full allowance is made for common mode (common cause) failures and it may be necessary to make an arbitrary allowance for the possibility of these.

## 18.5 Example

### Storage tank overfilling

The illustration here starts with consideration of a preincident situation looking at one very large storage tank. The numbers and design are based on tank T912 at Buncefield [3]. Unleaded petrol is delivered by pipeline from a remote site into a 7000-$m^3$ capacity tank at the rate typically in the range of 500 $m^3$/h rising to 900 $m^3$/h. An operator monitors the level in the tank using the automatic tank gauge (ATG) display on the control panel. The ATG also gives an audible alarm to alert the operator to indicate a high-level condition. There is an independent safety switch (LSHH) providing a visual and audible alarm in the control room indicating the ultimate high level was reached. This also initiates a trip to close valves on the relevant incoming pipelines.

The top event that is envisaged is overfilling of the tank leading to a loss of liquid from the vents in the floating roof. The tank has a standard bund designed to contain the full tank capacity. Taking this simplistic view of the top event and consequences the bow-tie diagram in Fig. 18.2 results.

This can be easily quantified starting with the number of filling operations per year, the estimates for the reliability of the prevention barriers based upon the design and testing of the ATG and the LSHH and the presence, training and workload of the operators. On the mitigation side the expected performance of the bund to retain all the liquid spillage should be checked. An ignition probability could be estimated and the consequences based on operator numbers in the affected area linked to a consequence analysis.

What happened at Buncefield was somewhat different. At 6.01 on the morning of Sunday 11 December 2005 there was a vapour cloud explosion at this large tank farm in Essex with an associated powerful blast and domino effects. The resultant fire engulfed 23 large fuel storage tanks on the site. These fires continued for 5 days and the plume of smoke spread over much of southern England. By good fortune there were no serious injuries or deaths. Large parts of the depot were destroyed and there was great disruption to surrounding motorways, and industry and over 2000 people evacuated from their homes.

The origin of the event was the overfilling of a large storage tank. Unleaded petrol was being delivered by pipeline into 7000-$m^3$ capacity tank from a remote site at the rate of initially of 550 $m^3/h$ rising to 890 $m^3/h$ from 05.50. It was later calculated that the tank would have been completely full by 05.20 although the level gauge recorded an unchanged reading from approximately 03.00 onwards. The safety systems intended to protect from overfilling did not function and an overflow resulted. Over 300 $m^3$ were spilt, most collecting as liquid in the tank bund but an estimated 10% vaporised and mixed with the surrounding air. At 05.38 a white mist was recorded by CCTV cameras thickening and spreading over the site, igniting 23 min later. The ignition point was probably either an on-site pump house or a generator house (or both).

What is clearly apparent is that the consequences envisaged in the simple bow-tie diagram were inadequate. The formation of a vapour cloud was omitted as this scenario had not been considered as sufficiently probable to be included in the safety case for the site—indeed it was not identified as the worst-case scenario. If it had been included then the RHS of the bow-tie—the event tree part—would have been greatly expanded and the consequences would have included the domino effects, offsite damage, the possibility of on-site and off-site injuries and fatalities, environmental effects and consequential losses and disruption.

Quantification of even the simple diagram in Fig. 18.2 would have shown inadequacies and the challenges of maintaining the barriers. The full diagram would have shown the need for major improvements in both the preventative barriers and the mitigation barriers.

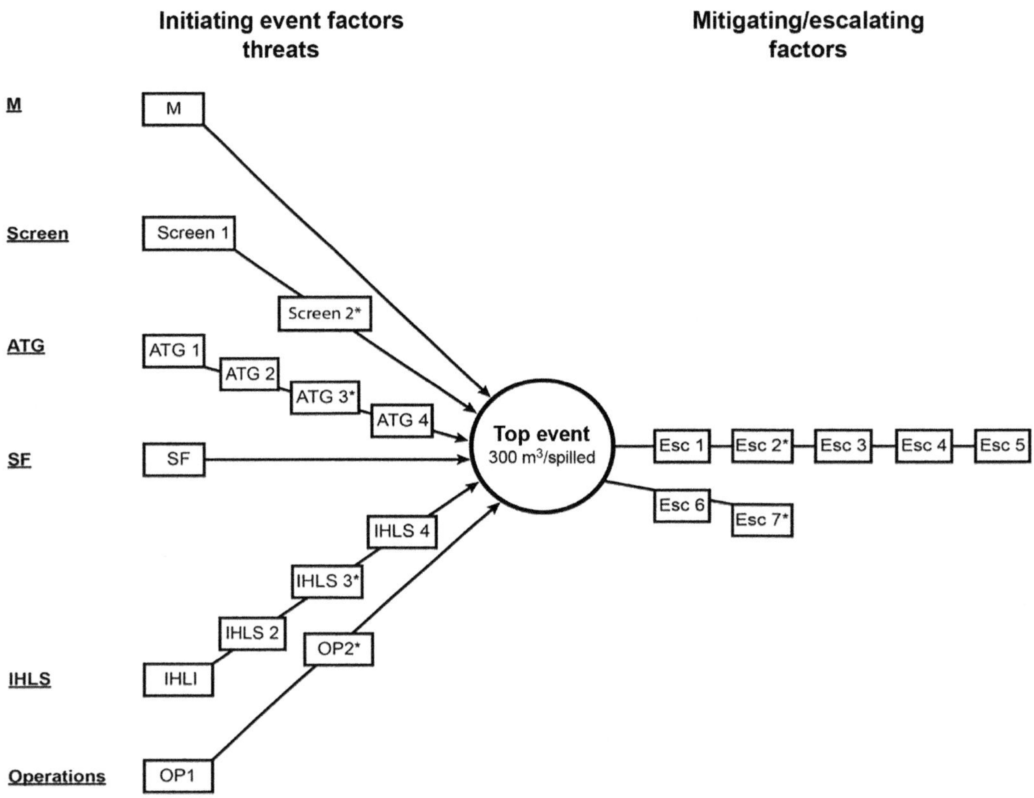


**FIG. 18.2** Bow-tie representation for Buncefield using the Why? Question and the official report (3).

**Notes and cross-references:**

| | |
|---|---|
| M | The operators were not given clear instructions on handling multisources so leading to confusion. See Chapter 16 (Vulnerability) |
| ATG 1 and 3 | See Failure Modes and Effects Analysis (Chapter 12) |
| ATG 2 | See Common Mode Failure in Fault Tree Analysis (Chapter 12) |
| Op 1 | See Chapter 16 (Vulnerability)—a classic form of Cognitive Dissonance |

**Abbreviations**

| | |
|---|---|
| HOSL | Hertfordshire Oil Storage Limited |
| ATG | Automatic Tank Gauging |
| IHLS | Independent High Level Switch |
| Hi | High level (warning level) |

Even this simple example shows how a clear picture can be created showing hazards and their interrelated causes and consequences. Amongst other uses they can be extremely helpful in learning from incidents, distinguishing the actuality from what might have been. They can also be used in many other areas of safety management, including training, operational risk assessments, maintenance programmes, human factor programmes and in audits [7].

## The Why? Question

| **Mechanical factor**<br>**Cause** | **Possible solution** | **Department** |
|---|---|---|
| The alarm did not sound | Check the alarm system (see below) | Instrument |
| Alarm did not operate | Establish a reliable function testing routine and take corrective actions | Instrument |
| The alarm was lost in the screen | Revise the screen layout | Instrument |

The actions for change, improvements (reducing the human factor) or monitoring are now defined and the responsible department defined. Such actions should be incorporated into corporate procedures or Standard Operating Procedures. If there is no monitoring process there is the risk of systematic erosion of controls. These issues should become evident in an Audit (Chapter 17).

In many ways the use of the 'why?' question is the start of a Root Cause Analysis as is shown in Fig. 18.2 (Table 18.1).

**Table 18.1** Narrative used in bow-tie (Fig. 18.2).

| Reference number as used in bow tie diagram, Fig. 18.2 | Dialogue or description of the event for the reference number in the bow-tie diagram, Fig. 18.2 |
|---|---|
| M | Multientry parcels (3) only one under HOSL control |
| IHLS 1 | New IHLS fitted |
| IHLS 2 | Operator not informed of a padlocked test override |
| IHLS 3 | High-Level Switch left overridden |
| IHLS 4 | IHLS Hi alarm did not sound in control centre |
| ATG 1 | ATG was very unreliable and stuck frequently |
| ATG 2 | The Hi and HiHi level alarms operated through ATG (common mode failure) |
| ATG 3 | Poor reliability of ATG was investigated but no remedial action taken |
| ATG 4 | Operators relied on the ATG system when receiving parcels |
| Screen 1 | Secondary information on visual screens was difficult to access and was behind other pages using the screen. This was not clear to the Operations Staff |
| Screen 2 | The operations group ignored this page/screen |
| SF | No proactive devices to shut down incoming parcels (human link only) |
| Op 1 | ATG showed that the level in T912 had stopped rising—the operator presumed that parcel was complete |
| Op 2 | Operators unaware that T912 was about to overflow |
| Top event | 300 $m^3$ petrol spilled out of the top of tank T912 into the bund. 10% is vapourised |
| Esc 1 | Vapour ignited when fire water pumps cut in |
| Esc 2 | This flashed back to the T912 bund |
| Esc 3 | T912 bund wall penetrations were sealed with soft mastic<br>This melted and increased the risk of fire spread and ground water contamination |
| Esc 4 | Escalation of fire (domino effect) |
| Esc 5 | Ground water contaminated with hydrocarbons and fire fighting foam products |
| Esc 6 | Delayed ignition of evaporated petrol with turbulence generators |
| Esc 7 | VCE |

*Notes*: *M*, the operators were not given clear instructions on handling multisources so leading to confusion. See Chapter 16 (Vulnerability); ATG 1 and 3, see example in Failure Modes and Effects Analysis—Chapter 12; ATG 2, See Common Mode Failure in Fault Tree Analysis (Chapter 13); Op 1, see Chapter 16 (Vulnerability)—a classic form of Cognitive Dissonance. *Abbreviations*: *HOSL*, Hertfordshire Oil Storage Limited; *ATG*, Automatic Tank Gauging; *IHLS*, Independent High-Level Switch; *Hi*, High level (warning level).

## Dialogue for Buncefield bow-tie

The asterisk * is the likely start and end points of the bow-tie when drawn on its own. However, the use of the 'why?' question encourages the development of the branch into potentially concealed but significant initiators. This is represented by Fig. 18.2, which is a Root Cause Analysis in bow-tie format.

# References

[1] CCPS, Bow-Ties in Risk Management: A Concept Book for Process Safety, 2018.

[2] M. Manton, M. Johnson, R. Pitlado, et al., Standardisation of Bow-Tie Methodology and Terminology via a CCPS/EI Book, IChemE, 2017. Hazards 27 Conference Paper.

[3] Buncefield Major Incident Investigation Board, The Buncefield Incident 11 December 2005, (Final Report), HSE Books, 2008.

[4] G. Guiochon, L. Jacob, On the Catastrophic Explosion of the AZF Plant in Toulouse, CCPS—Center for Chemical Process Safety, September 21, 2001.

[5] Manton, et al., Representing human factors in bow-ties as per the new CCPS-EI Book, in: Hazards 27, IChemE Rugby, 2017.

[6] Plant Ageing: Management of Equipment Containing Hazardous Fluids or Pressure (HSE Research Report RR509), 2006.

[7] G. Pilkington, How to use bow-tie diagrams, Chem. Eng. (April 2017) 48.

# 19

# Process hazard review (PHR)

Chapter outline



This chapter examines an existing plant and uses operational experience to identify potential operational and safety issues created by changes in procedures and design intent through modifications. While it is a HAZARD Identification technique, it does not fall logically into any of the previous chapters so is included as the last chapter.

## 19.1 Definition

PHR is a structured review of an existing process using the existing process or plant knowledge base.

The intent is to identify drifts from design intents through modifications and previously unidentified operational issues and the possible impact of new regulations.

## 19.2 Description

PHR is used on an existing plant where there will already be some operational experience; it uses some of the elements of a Safety Audit and some of a HAZOP study. It draws on the experience to be found in historic incident reports, plant records and in the knowledge base of the operating staff so as to identify areas where there may be requirement for an improvement in the safety features (hardware or software) to satisfy new or existing regulations or as a result of process changes/enhancements [1]. It is a development of a previous technique by that name developed within ICI.

As with an audit, the scope of the PHR study must be clearly defined and the information required must be identified. The team is small and is representative of the knowledge

A Guide to Hazard Identification Methods. https://doi.org/10.1016/B978-0-12-819543-7.00019-7

and responsibility/experience on the plant and should include suitable specialists. The analysis is carried out in a structured approach on small but discrete sections of the plant, reviewing the existing knowledge to identify significant hazards and then examining the existing levels of protection, procedure, hardware or other in place. The team may use some of the HAZID approach (Chapter 5), the Checklist approach (Chapter 10) or a 'What-If?' approach (Chapter 11) to investigate hitherto unrecognised hazards such as potential sources of loss of containment, places where personnel may be trapped or equipment (such as shutdown systems), which has a poor performance.

The problem, as with any 'identification process', is for the Facilitator to gain the confidence of the team who will be highly oriented to the production process. The first few minutes are critical. It would be useful to have a series of prompts which from experience could be over 20 and may include such as:

- What are the good features of your plant? (Start with the positive approach.)
- What are the weak features of your plant?
- What parts would you like changed? (This allows the term to air its own feelings.)
- What equipment gives you operational difficulties?
- What systems seem to be less reliable than you would wish?
- Are there any process shut down features that cannot be properly tested on line?

This list is probably best developed by each Facilitator to fit in with his/her own style. See also the hints on body language in Chapter 4 (HAZOP) and Chapter 17 (Audits).

The report of the PHR will normally be given to a senior staff representative in the form of a series of actions, which the local management can then resolve. It is essential that the report is objective and, like an audit, is suitably focused and structured. It should aim to avoid trivia and fine detail but also try to focus on the main issues, some of which may be generic. Actions should be prioritised if there are a large number.

A possible work sheet is shown in Table 19.1 but again the Facilitator may have his/her own approach.

**Table 19.1** Typical PHR report table

| Reference number | Issue identified | Concerns | Possible actions | Ranking |
|---|---|---|---|---|
| 1 | Poor access to valve ABC | Possible back injury | Consider:<br>1. Valve extension<br>2. Chain operator<br>3. Access platform | M |

Please treat this as an indicative issue. It is virtually impossible to generate a real situation without knowledge of the process.

## 19.3 Resource requirements

The requirements are for no more than items in normal use such as plant drawings lay out, P&ID, operating instructions, maintenance records and, whenever it is available, operating experience on this and other similar plants.

The team membership would not normally exceed five persons, including as a priority operations and maintenance personnel and specialists as necessary.

The duration will be dependent on the scope and plant size but would be typically about 5 days.

## 19.4 Timing

The study can be carried out at any time during normal operation. It is most beneficial after some years of operation or following a significant change in regulation or design intent.

In the latter case of new regulations it would be more appropriate to have a team of about 3 who would look at the implications of any new regulation and then to make recommendations for any actions.

## 19.5 Advantages, disadvantages and uncertainties

### Advantages

- The study can be carried out on a process that has not been subjected to traditional rigorous studies during its development. However it should not be used lightly on major hazard processes.
- The study can be used following systematic changes due to process enhancements and regulatory changes.
- The study can have an element of independence.
- The study uses the existing process knowledge base and historic experience of that type of process.
- The study can give a quick assessment of future study needs such as a new HAZOP.
- The study can identify problems that may not be identified by the original HAZOP study.
- The study requires considerably less effort than a full HAZOP study.

### Disadvantages

- The study will not be in as detail as a traditional HAZOP study.
- The report has to be carefully thought through to avoid dissipation of effort.

### Uncertainties

The depth of the depth and detail in the knowledge base within the operations group may influence the results of the study. The team has to be chosen with care to show the depth of knowledge.

## 19.6 Use of PHR

PHR is potentially very flexible. The scope will influence the manner in which it is handled. Uses outside the more obvious areas have included such as identifying areas of poor availability leading to potential loss of production and routes to improved Standard Operating Instructions (SOP).

## Further reading

[1] ABB Consulting, https://new.abb.com/.

# Appendices

Many of the illustrative examples when small have been incorporated with the topic itself, but the following are more expansive and require particular attention.

Appendix A shows the techniques for a nonprocess situation.
Appendix B shows the full range of techniques for a continuous process.
Appendix C shows the range of techniques for a batch process.

# Appendix A

## An example of hazard studies during a modification used in a nonchemical environment

### A1.1 The problem

The task is to repair a large underground drain carrying the site effluent to an effluent treatment plant. The section involved is shown in Fig. A1.1.

The concrete at one of the manholes is crumbling and must be built up in an appropriate manner. The **Inherently Safe** way of carrying out this repair would be to shut down the site. However there are potential hazards, which must not be overlooked, during the shutdown and the start-up of the site. It can be taken that this option is the final fall back position if the 'on line repair' is considered to carry an excessive risk.

#### Preliminary hazard reviews, prior to budget approval (PHS 0 and 1)

The first stage of the **process hazard analysis (PHS stage 0),** based around a brainstorming session, generated possible solutions ranging from building a new parallel drain, bypassing the damaged section with a new drain section to repairing the damage in some manner. It was noted that the first two would involve a 'break in' and so present a significant risk of loss of containment. Due consideration was also given to the '**Inherently Safe solution**', i.e. to shut down the whole site. This was rejected not only due to the cost of lost production but also for the risks associated with the shutdown and start-up of the site. The method finally selected was to install a temporary bypass to the affected section, to plug the damaged section and then rebuild as necessary.

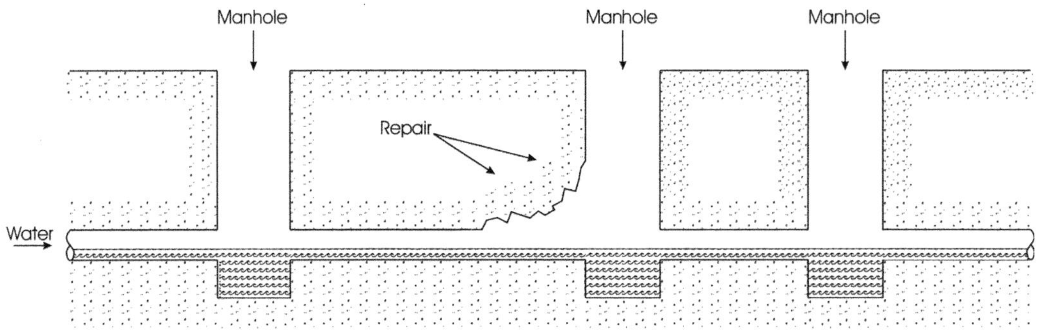


**FIG. A1.1** Damaged drain section, including proposed temporary bypass. *Courtesy of IChemE. (Extract from Practical Risk Assessment IChemE Rugby, no longer in press.)*

With a method now selected, the concept stage hazard review (**PHS 1**) concentrates on seeking the potential SHE hazards. A combination of **what-if** and **checklists** is used. The cause of the damage must be determined so that a compatible repair material can be selected and potential problems with both the repair work and with the use of the temporary bypass are identified. Amongst the issues for the **what-if** study are:

- 'What if the liquids are too hot?'
- 'What if effluent from plant A reacted with plant B to create an aggressive compound?'
- 'What if the plugs perish?'
- 'What if the pump corrodes?'
- How is the plug fitted at the start of the operation and removed at the end?
- What if there is a release of 'flammables' near the work site?

The **checklist**, in addition to a routine listing of material properties and hazards, will include:

- Aggressive compounds.
- High temperature (problems with net positive suction head).
- Other reactions/reactants.

This study leads to confirmation of the cause of the damage and allows specification of the repair method, using an epoxy resin. The procedure will be to install the pump and temporary bypass, then to plug the up- and downstream outlets of the drain, to clean and purge the isolated section, put in any necessary support to the damaged area and then to start the repair work.

## Design stage reviews (PHS studies 2 and 3)

The **front-end engineering design (FEED) study** will follow very closely the completion of the concept selection, which will have included the selection of the type of epoxy resin to be used. It will address the risks to the repair team and also develop the **Basis of Design (BoD)** for the plugs and pump out process. A blend of **checklists** and **what-if** studies will be used, with **HAZOP** if appropriate.

The **checklist** will include the following items:

- Means of escape.
- Personal protective equipment.
- Emergency planning.
- Rate of flow of effluent.
- Contaminants in effluent and impact on the repairs/process. This will indicate the need, or otherwise for 'Air Tests' during the task.
- Duration of the task.
- Toxicology of contaminants.
- Plug fitting—remote or by entry to the sewer.
- Effect of the contaminants on humans and the repair compound.
- Effect of the repair compound on humans.
- Compatibility of the work and the immediate environment.
- Impact on road access.
- Action if there are any flammable releases near the work site.

The **what-if** study will parallel and extend the checklist work. A selection of the questions is shown below.

- What if there is a rain deluge? (High flow)
- What if the running pump fails? (No flow)
- What if the plug fails or is dislodged? (More flow in a No-flow line)
- What if there is a sudden draining on a plant on the site? (More than)
- What if the manhole collapsed?
- What if the plugs leak?
- What if there is a fume build up in the drain?

A **HAZOP** study may also be used as a 'Preliminary Study' for the effluent and pump out system. Deviations considered will include:

- <u>No flow</u> of effluent. How does the pump handle changing rates? How quickly can a spare pump be started up? Will the pump lose suction if the temperature is too high?
- <u>High/low flow</u> of effluent. What control system is needed and is a 'minimum flow' recycle needed (to avoid overheating)?
- <u>High</u> effluent <u>temperature</u>. Cause? Effect (Net positive suction head requirements)? Warning? Action?
- <u>High temperature</u> in the pump. Minimum flow needed? See above.
- <u>Higher concentration</u> of 'Compound X'. Cause? Effect? Warning? Air test requirements? Actions/materials that require careful specification?
- <u>No flow</u> of diesel fuel. Where are reserves of fuel to be stored? Where can emergency supplies be obtained, e.g. at midnight? What is the environmental implication of storage outside a designated area?
- <u>High flow</u> of effluent due to a process upset or rainstorm.

- High flow of fuel (as above).
- No flow of fuel, that is, the pump is shut down.
- Other than diesel, that is, petrol and not diesel fuel.

The final results from these studies will include:

- A listing of the potential contaminants in the drain system.
- The toxicological properties and by-product properties of the contaminants.
- A peak flow of effluent.
- A means of fitting/removing the plugs. This could be by a special, controlled entry, or it could be by remote control. (Such devices are available.)
- Preliminary PPE, which is compatible with the contaminants.
- Initial support plan for the manhole.
- Initial emergency plan.
- Road access impairment (and any implications for emergency or other vehicles).

The Basis of Design (BoD) can now be written. This would include flows of fluids and an outline specification of the capacity of the pumps and the material of construction. Likewise the materials for the drain plug would either be specified or the materials' specification would be drawn up with the temperatures that must be tolerated and the chemicals that it might be exposed to.

## PHS Study 3 detailed design

The detailed design can be developed. The reinforcing/support to prevent collapse of the manhole can be designed and a means of fitting it without entry to the manhole can be established (this may only be wooden props but it could be more sophisticated). The PPE can be fully specified and purchased. The specification, number and type of pumps, (including materials specification), can be drawn up, with a reliability or performance criterion, for purchase and the pump bunding (against fuel spillage) and location finalised. Fuel stores can also be finalised with any requirements for fire fighting. The design of the pump level control can be finalised. Finally the specification for the plugs (including the materials specification) can be drawn up for purchase.

Plans may also be drawn up for the 'control and monitoring of effluents' during this task and a small control centre established.

A checklist can be used to ensure that the design reflects the BoD. It would be prudent to carry out a further HAZOP in some detail, but for such a project it is unlikely that it will reveal any new major issues.

## Construction and pre-start-up studies (PHS studies 4 and 5)

The construction phase requires the drawing together of the main elements of work. It will be more a case of ensuring that everything is tested and in the correct place. It will not a true major 'construction' exercise other that the support for the manhole, the pump piping, the fuel store (which may be no more than several jerry cans), the escape frame and also the writing of the emergency procedures and their practice.

The construction of the support for the manhole will require a test criterion and inspection. The PPE must be obtained and kept at the work site, for example to cope with the fumes from the effluent, which are irritant and the fumes released from the epoxy resin, which are mildly narcotic. The pump bays and bunding (against spills of fuel) and the fuel store must be built to an appropriate standard.

Emergency procedures will be written for situations such as:

- Leakage of the plug.
- Fuel losses.
- Escape from the manhole.
- Sudden storm.
- Loss of pumping capacity/control.
- Effluent monitoring and control.
- Flammable releases (if appropriate).

This could be classified as part of the design as some extra features arising from the HAZOP may have to be incorporated. The preparation of the Emergency Procedures may come at the interface between the two studies.

The pre-start-up work will include a check on the training requirements for the various team members, including rehearsal of the possible emergencies. The repair team must be given a clear description of the work to be carried out and the emergency actions.

It is quite likely that this phase will adopt a **checklist** approach (punch list or reservation list) and will cover such items as:

- PPE in place and tested?
- Resin purchased and in the store?
- Plugs ready to fit and tested?
- Pumps (and piping) in place in a bunded area and test run?
- Air movers available?
- Supports for the manhole designed, fabricated and ready to be installed?
- Escape harnesses and lifting frame in place and certified as adequate?
- Fuel stored and emergency supplies identified? Bunding specified and isolated from the storm drains? Fire extinguishers (foam type) ready and clearly identified?
- Stand-by person trained?
- Repair team trained?
- Emergency procedures written, practised and ready to be put into operation?
- Road access is properly controlled/monitored?

A site tour should be carried out in readiness for the **Permit to Work (PtW)**. Air tests can now be carried out as required from the survey in the **FEED**. This will include cold work and entry. Before the permit is issued the work site would be inspected and all operating units on the site made aware that the drain system was to be put under repair for a number of days. The PPE may include 'green suit', boots, gloves, full-face visor, Breathing Air, forced ventilation of the manhole, harness, lifting frame and 'stand by man'. All of these requirements were determined at the earlier stages of the project and can now be put into a **checklist**. In many ways the PtW will be a reflection of all of the Hazard Identification and could be derived from a checklist that has been developed during the Concept Selection, the FEED, The Detailed Design and then Final Construction.

A final check on the 'effluent control centre' would be carried out and site warning that the repairs were about to start would be given.

Routine air tests may be appropriate during the work. These tests should be taken a low point in the drain pit, not at the top. When the work begins the first task will be to establish the by-pass route and then to fit the plugs.

## Postcompletion (PHS 6 and beyond)

A **checklist** will be prepared to cover the removal of the plugs and the temporary bypass and return to normal use of the drain. In addition, periodic monitoring of the whole drain will be carried out to check for any signs of recurrence of the problem.

# Appendix B
## Example of application of hazard identification techniques during the life cycle of a large continuous process

## B1.1 Example plant

A new olefin plant is to be built on a Greenfield site. The process is largely fixed, involving the cracking at 850°C of ethane or, less likely, a propane/butane/naphtha feedstock. The products of the cracking process will be compressed, cooled and dried. After refrigeration the products will be separated by low-temperature distillation. A waste heat, steam-raising facility is to be used to drive the compressors.

There are many issues that have to be addressed during the design; indeed are too many to include in a worked example such as this. They will include the transport of feedstock and product, the impact of the installation on the local services, including the emergency and design services and, of course, the environmental impact. However, to develop the illustration of Hazard Identification tools, two topics have been selected for attention.

1. The $CO_2$ and $H_2S$ removal system could be by either caustic soda or triethanolamine (gas sweetening). The caustic wash system produces an alkaline effluent stream, while the triethanolamine wash requires regeneration and combustion of the $H_2S$ in a flare or equivalent.
2. The acetylene removal could be done by selective hydrogenation over a catalyst. Two options are available, either by hydrogenation of the raw gases prior to gas separation (front-end) or by hydrogenation of the ethane/ethylene stream (back-end). The front-end hydrogenation is less selective but will also remove methylacetylene, which is unstable and is otherwise rejected in a propane stream. The back-end hydrogenation is 'inherently less safe' but gives a better yield through a slightly lower conversion of ethylene to ethane. At higher concentrations the methylacetylene in the waste propane stream can spontaneously detonate. Acetylene in high concentrations and ethylene at high temperature and pressure can also detonate. The control of the concentrations and other parameters of these streams is very important but is not insurmountable.

## Preliminary hazard reviews, prior to budget approval (PHS 0 and 1)

The first stage of the **process hazard analysis (PHA stage 0)** will include a search for inherently safer approaches, covering both environmental hazards and the minimisation of the working inventories, especially in the refrigeration and gas separation systems. Both **checklist** and **what-if** methods would be used at this stage. Decisions will be made on the options outlined above for gas sweetening ($CO_2$ and $H_2S$ removal) and the acetylene-handling method. Environmental issues include spills, wastes from the $H_2S$ removal or the impact of the spent caustic and associated polymers if the caustic route is chosen. This phase will take only a few weeks review and justification for one person.

At the conceptual design stage, the remaining components in the system will be defined. As the processes of thermal cracking and gas separation are well established there are likely to be few difficulties here. The deciding factors in furnace design are likely to be the specific needs of the downstream product spectrum and nature and costs of the feedstock. The separation and acetylene hydrogenation units will be part of a design house package with the final decisions based on a bid analysis, which is not strongly influenced by the SHE features, although these need to be identified here to ensure the designs provide satisfactory coverage. In the **PHA study 1 (conceptual design)**, **checklist** and **what-if** approaches will be used. Many aspects must be considered, including layout and location of the control room (against a VCE), fire protection, general SHE issues, water supplies, operating personnel requirements, timing of delivery of key pieces of equipment, etc. The key issues for the **preliminary risk assessment** should be identified and **fault tree (FTA)** and **event tree analysis (ETA)** will be used to clarify the causes and likely frequencies for the main hazards. An early **layer of protection analysis (LOPA)** may be carried out to indicate the level of shutdown system that will be required.

At the end of this stage a 'preferred contractor', based on best overall rate of return on investment, will be chosen to develop the project. The bid/selection process may take some months.

## Design stage reviews (PHS studies 2 and 3)

The **front-end engineering design (FEED) study** will be incorporated into the early design process, which itself will follow very closely after completion of the concept selection. With the key issues already identified, **what-if** and **checklist** studies will be used to assist the identification of second-level risks. In turn these will be assessed using **ETA** and **FTA**. As equipment bids are received from the selected suppliers **failure modes and effects (and criticality) analysis (FME(C) A)** may be used to identify the most robust/reliable items.

During the development of the P&IDs use may be made of **LOPA** to assist in deciding the initial safety integrity levels (SILs) of the safety instrumented systems (SIS). A **checklist**-led review of the early P&IDS will be done with the aim of identifying the more obvious errors and so reduce the need for design change requests later in the project development. **Checklist** and **what-if** approaches will be used to ensure consideration of the more general issues, including layout, noise, visual impact, effluent sources and disposal, sectional isolation, means of escape and maintenance access. This phase of the project may last for some months, but the SHE effort will be only a small percentage of the total work.

As the detailed design emerges the PHA 3 studies will be initiated. A major effort will be put into the **HAZOP study**. This is vital to the validation and finalisation of the P&IDs and is likely to take a team of 6 persons several weeks. During this phase some of the process piping may be prefabricated and so there is the potential for a conflict of priorities. The more detailed definition of the design allows refinement of the **LOPA** and associated **ETA** and **FTA**. This work will be needed for the preparation of the **Safety Case**. Whilst not a fundamental part of this document these will be augmented by consequence analyses into a full Quantified Risk Assessment.

Other SHE studies, using **checklist** approaches supplemented by the **HAZOP**, will be needed to analyse/assess the performance of the relief and blow down system. Hazardous area classification can be based on a '**rules**' approach and **checklists** may be used to consider means of escape, mechanical handling and emergency response.

The SHE effort will run throughout the detailed design work over at least a year and it will form a significant part of the total effort when the safety case work is included. As indicated previously, there will be overlap with prefabrication and the start of construction.

## Construction and precommissioning studies (PHS 4 and 5)

A detailed **checklist-based audit** of the plant will be done, based on the 'as built P&IDs'; it will cover not only the equipment but also the materials of construction. A check will be made against the list of 'construction aids', which must be removed, e.g. pins in spring supports, balanced bellows relief valve plugs. **Checklist** reviews of procedures and document systems will be carried out to ensure the plant is safe to commission and operate. This includes a review of all actions from the **previous hazard studies**, especially the **HAZOP** study, to ensure they have been implemented. Once construction is well underway this SHE effort will be a very significant fraction of the process effort.

The precommissioning review will use **checklists** to cover key areas such as training, SIS testing, cleanliness, purging, pressure testing, drying standards and emergency practices.

## Post-start-up studies (PHS 6 and beyond)

The overall review of the 'lessons learned' can benefit from a **checklist** approach. It should include within its scope the technology, project management, construction successes and failures and problems identified by the HAZOP of the actual start-up.

A sound 'management of change' system will be needed to ensure an appropriate level of review for all proposed changes to the plant, the operating procedures or the control systems. An initial review using a **checklist** or **what-If analysis** will always be used. For higher-risk changes the original **HAZOP** will be reviewed and an updated version prepared based on the implications of the proposed change.

At regular intervals, for example every 5 years, a review will be carried out focussing on the changes made since plant start-up. This will include **checklist** review of the safety management systems, use of site **audits** and a revalidation of the previous **HAZOP** study. Routine application of **PHR** may be appropriate as an assurance that SHE standards are being maintained.

As decommissioning and demolition is a one-off event it will be planned after a predevised corporate **checklist** has been used to review the intended operation.

## B1.2 Some illustrations of study details and outcomes

As a means to illustrating the operation of the techniques some short (and incomplete) examples are given. It is essential that these are treated as indicative and not as a full and complete solution. However they do indicate both the comprehensive and complementary manner in which the different techniques can produce good coverage of SHE problems when correctly applied.

### Preliminary hazard reviews, prior to budget approval (PHS 0 and 1)

*Checklists for PHS 0*

The checklists would include, inter alia:

**Environmental impact**

Noise
Odour
Visual
Effluent—liquid or gaseous or solid

**Inherent safety**

Intensify
Attenuate
Substitute
Segregate incompatible sections of the process
Toxic or hazardous properties of products/by-products
Etc.

**Acetylene removal**

Front-end
Back-end
Risk ranking

**$H_2S$ removal**

Safety
Environment
Wastes and disposal
By-products

*What-if at PHS 0*

The 'what-if…?' questions cannot be formulated in totality for this illustration but would certainly include:

there is an environmental impact?
there is a significant difference in inherent safety between alternatives
the acetylene converter runs away?
there are by-products from the $H_2S$ removal?
there is a gas leakage in the Gas Separation Area leading to a major fire or a vapour cloud explosion?

*PHS 1*

The **checklists** and the **what-if** questions are likely to be of a similar style and content. The preconcept (PHS 0) studies will have pointed up the key issues, which must now be addressed.

The use of the **Relative Ranking** and **RAST** coupled with some **FTA** and **ETA** should give guidance on the benefits to be derived from enhanced safety/environmental features. More particularly, in this case, **LOPA** may be applied to assess the likely level of protection to be applied to the acetylene converter.

The key issues, which will come out of this analysis, are the $H_2S$ removal route and the acetylene removal/conversion. Others, which may come out of the analysis, are the segregation of the pyrolysis furnaces and the gas separation as well as the control room location, the impact on/from neighbouring plants and the impact on the public off the site from noise, visual impact from flare stack and the damage potential from an on site VCE.

## Design stage reviews (PHS studies 2 and 3)

The key item, which has to be addressed in the **front-end engineering design (FEED)** study, is the integrity of the converter shutdown system (back-end only). This will involve **LOPA** and progress onto a detailed **FTA** using a corporate risk criterion. The layout on the macrodetail will have to address the segregation of incompatible systems such as the control centre and large sources of hydrocarbon. A second set of questions could be derived to examine the lay-out at microdetail.

**Checklists will include**:

Access for maintenance
Loss of containment due to impact, dropped or swinging loads, physical
Impact from external sources.
Means of escape.

*What-if study*

The what-if questions cannot be formulated in totality but will include:

there is poor maintenance access?
there is a physical impact causing loss of containment?
there are poor means of escape?

The detailed design will take account of many design procedures and standards. The PHA 3 studies will include **Hazardous Area Classification** and a **Relief and Blow down Review**. The use of **HAZOP** would also confirm the problems associated with the $H_2S$ removal and the acetylene converter. A part of the HAZOP study report follows (Table B1.1).

**$H_2S$ removal**

The HAZOP study will show whether the LOPA was sufficient to determine the SIL of the converter protective system; if not, more detail will be required from a further FTA and/or ETA.

## Construction and precommissioning studies (PHS 4 and 5)

The **checklist** would include questions such as:

Built as intended?
Materials
P&IDs checked in 'as built'
Any errors incorporated by deviation from the design intent?
Are all the actions from previous hazard studies signed off?

Questions on the pre-start-up **checklist** would include:

All modifications complete?
Training complete?
Operating instructions complete?
Trip tests complete?
Preparations carried out fully (cleanliness, drying and purging)?

## Post-start-up studies (PHS 6 and beyond)

The **checklist** for demolition would cover, inter alia:

Decontamination—plant and soils.
Integrity of the walkways and lifting points.
Order of removal.
Equipment for sale to second party and equipment for scrap and standard of cleanliness of sale items.
Demolition safety case.
Documentation for the Environment Agency
Documentation for the demolition contractor.

**Table B1.1** Extract from HAZOP report.

| No | Parameter | Guide word or deviation | Description | Cause | Effect | Protective systems | Actions | On |
|---|---|---|---|---|---|---|---|---|
| 1 | Flow | No | No flow of ethylene/ ethane Stream ($C_2$) | $C_2$ Control valve fails closed | Converter enters an exothermic reaction leading to detonation of the contents | Low-flow trip FTL123 LOPA analysis already carried out | A1. Review the level of protection required to achieve safety criteria | AB & CD |
| 2 | Flow | No | No flow of hydrogen ($H_2$) | $H_2$ Control valve fails closed | Product off spec due to acetylene | Low-flow alarm FAL 124 | A2. Determine if a system involving a dump valve to flare is required | CD |
| 3 | Flow | High | Excess $H_2$ over $C_2$ | $H_2$ Control valve fails open or the $C_2/H_2$ ratio controller fails | Converter enters an exothermic reaction leading to a potential detonation | None | A3. Review the level of protection required to achieve the safety criteria<br>See A1 | AB & CD |
| 4 | Temperature in the converter | High or higher | As above | As above or due to $H_2$ CV leakage at a shutdown | As above | High-temperature trips in the reactor bed | A4. Review the need for ratio alarms on the process streams<br>A5. Review the need for a second ESD valve in the $H_2$ stream | EF<br><br>EF |
| 5 | Temperature in the converter | Low or lower | Converter cools and reaction stops | $H_2$ Control valve fails closed or the $C_2/H_2$ ratio controller fails to increase the $C_2$ stream flow.<br>$C_2$ control valve fails open. | Product off spec due to acetylene | None | A6. Review the need for a high flow alarm on the $C_2$ stream and/or a ratio discrepancy alarm. Also review the need for a low-converter exit-temperature alarm | CD |
| 6 | Composition | High or higher | Converter conditions non ideal. Low conversion of acetylene | Deviation in any of the control parameters | Product off spec due to acetylene | Exit analyser AN125 | A7. Review the response time of the analyser and consider if there should be some deviation alarm in the bed temperature measurements | AC |
| 7 | Flow | As well as | Flow of ethane and ethylene contaminated with water | Conversion of any Carbon Monoxide in the $H_2$ to methane and water (methanation) | Hydrate formation in $C_2$ Splitter | None | A8. Review the need for analysis of the $H_2$ stream either by sample or by analyser | BD |
| 8 | Flow | As well as | Flow of ethane and ethylene contaminated with 'green oil' | Reaction of acetylene to form benzene derivatives | Fouling in $C_2$ Splitter | None | A9. Review means of removal of frozen 'green oil' with methanol injection | AC |

It will be noted that the issues surrounding the converter will be identified by a number of the techniques. It should also be noted that the detail will vary according to the technique used and the time of use. Thus the **checklist** and **what-if** studies will identify the problem and **LOPA** will give some indicative estimate of the safety integrity level (SIL) required of the protective system. However, it is the **HAZOP** and the **FTA** that will give the final and definitive answers to the SIL requirement.

It is possible that some modifications will be needed, for example, process optimisation. A **safety review** will be carried out for every change, using a **checklist** to decide what level of assessment and approval is needed. This may vary from a **simple risk assessment** made by a small team up to a repeat **HAZOP** study of a section of the modified plant or process. It is important to decide if the **SIL** requirement has changed. A review of the changes proposed, at up to Operations Director level, is required before the changes can be implemented.

Clearly the Hazard Identification techniques must be used in a manner that is most likely to produce the relevant information for current phase of the project. Timely use of an appropriate method will define the potential issues, so they can be resolved at the most appropriate phase of the project.

# Appendix C
## Example of application of hazard identification techniques during the life cycle of a batch process

### C1.1 The project

This example is based on the introduction of a new process into adapted plant in a company (TPW Ltd) that runs many different production campaigns each year. TPW is a long-established general chemical manufacturing company, employing over 2000 people at several sites in the United Kingdom. It has a good safety record and has a central safety group to oversee the hazard studies and to provide trained leaders for all process hazard studies on new plant or processes.

The site that will house the new process is on the outskirts of a northern industrial city and employs 350 people. It lies between a river and a major road and on one side is close to an old housing estate. There are a number of continuous processes and several general-purpose batch units on site. If the new process is approved for production it will be within one of the batch units after modification to meet the requirements of this process.

TPW has been approached by an existing customer and invited to bid as the preferred supplier for production of a product that will be new to both the client and TPW. The client has a suggested route that has been tested to 1 kg batch size in their own technical facility. The initial production requirement is for 20 Te, which would require a campaign of repeated batches over several weeks. If the first campaign is successful there are likely to be repeat orders for even larger amounts.

The process involves a number of relatively routine operations, particularly in preparation for the main reaction steps and the final product treatment. The unknown stages are the two main reaction steps. In the first of these, the sodium salt of an organic reactant, A, is formed by adding caustic soda solution to A in a large reaction vessel. This reaction is mildly exothermic; a slight excess of caustic is used to maintain the pH around 11. The salt is then reacted with a second organic material, B, added at a controlled rate from a measure vessel. This reaction is very exothermic, and cooling is required. To obtain a high-quality product the reaction needs to be carried at between 55 and 60°C.

When a new product proposal is evaluated by TPW there are several areas where objections may prevent the proposal being taken forward. SHE issues form one of these areas. Others are technical feasibility given the available plant and the financial viability of the project. The project will be developed by several groups within the company, working in parallel, with a number of 'gates' at which the business team must give approval to the continued consideration of the project. Failure at a gate may be due to the identification of critical economic, technical or SHE problems. As always, there is 'time to market' pressure on the groups to complete the work as quickly as possible.

## Preliminary hazard studies (PHS 0–2)

The first, preliminary review is based on a **checklist** but equally could include **Relative Ranking** and **RAST**. It covers the technical feasibility of the process, overall regulatory issues, availability of suitable plant, and looks for potential SHE issues and chemical hazards. This review is short, typically taking one person a day. The first gate follows this review.

Before the next gate, sufficient information must be available for the business team to be confident that the company can safely produce the material, meeting all regulatory conditions. It should also be possible at that stage to give the client ballpark figures on times and cost. The necessary data will be generated by three teams, working in parallel and each having a specific role, but collaborating closely and sharing data. The SHE group, using their **Initial Assessment Checklist**, will seek to identify any requirements under the companies COMAH Safety Report, Hazardous Substances Consent, planning regulations and chemical registration for materials on site. They must also confirm that IPC/IPPC permits covering all emissions (solid, liquid, gas) should be obtainable within the planned timescale for the process development. The Technical Operations group will identify the plant to be used, determine the main modifications required and arrive at a rough estimate of the cost. They will consider a number of safety-related issues, including materials of construction, instrumentation and control, as well as unit isolation and relief. The Development Team will carry out a literature search on the material and the process, including the listing of the obvious chemical hazards, and then run laboratory checks. A small-scale laboratory preparation will be done to prove the route and to check the reaction conditions. Reaction heat flows and gas flows are then measured using reaction calorimetry (at 500-g scale). The stability of all reactants, intermediates and products is checked using differential scanning calorimetry (at mg scale) and differential thermal analysis (at g scale). If successful, further runs at up to kg scale will be done to define the process, leading to a process outline for the plant operation. An outline process description is now written, justifying the route and including safety assessments for all materials. Waste production and emissions will be identified, roughly quantified and disposal routes outlined. The Team must be confident they can complete any new **COSHH assessments** that may be needed. When the three groups have completed their studies, a Business Team review decides if the project can proceed.

In the next stage the Technical Group moves towards development of a detailed design, including the operational methodology. The immediately available information includes the mass balance, a process flow diagram, process description, the process chemistry and the expected wastes, and Safety Data Sheets for all the materials expected to be isolated (i.e. where there is the possibility of exposure). The first design stage review, including the **Hazard Study 2 (PHS 2),** is now done. Normally a team of 5 or 6 is assembled, led by an independent chairman and involving the process manager designate, process engineer, chemist and a mechanical engineer. A **what-if** approach is used, aided by a set of **key-words**. The keyword list includes flammability, chemicals, health issues, environment, human factors, relief, drains, worst-case scenario, services, control, etc. This study will take one or two ½ day sessions and will generate a number of queries for both the process chemist to investigate and the engineering team to resolve.

Several of these areas are then individually examined in more depth. The **worst-case scenario(s)** must be identified. This is done by a **what-if** study aided by a list of more common causes, such as more of or less of each chemical (the larger of one unit of addition or $\pm 20\%$); faster addition, agitation (none, 50%, more); loss of cooling; temperature or pressure variations. The evaluation of consequences considers what happens immediately, what happens in the next vessel and also questions the final fate of the upset batch. This study will be supplemented by **human factors analysis** that is intended to ensure that all reasonably foreseeable mal-operations are considered. As part of this analysis a small group, including an operator and a plant manager, 'walk through' the complete process, step by step, asking what can be done wrongly. As in a HAZOP, each deviation suggested is analysed for its consequences and to see if the existing protection is adequate. The **prompt list (guidewords)** used in this study includes not, early, late, twice and wrong order. These studies should identify the most important and most probable scenarios. If necessary **ETA** will be used to fully explore consequences and **FTA** can be used to ensure that all causes of a particular outcome have been identified.

Some of the identified scenarios will be investigated experimentally by reaction calorimetry so that vessel venting requirements or other protective measures can be specified. It is possible for a change of process to result from these studies, for example, batch size, rates of addition or operating temperature range. At the conclusion of this work the **preliminary CHA** (Chemical Hazard Analysis) can be written, including a definition of the **chemical basis of safety** for the process; this may, for example, be guaranteed agitation, a maximum addition rate for a key reactant, maximum and minimum operating temperatures, etc. Meanwhile the SHE group will have done further work checking whether the introduction of this process might affect existing **COMAH and IPPC approvals** and the Technical Group will have produced a detailed design, including a set of P&IDs and the associated design data. A sanction package is now put to the Business Team, seeking approval to proceed towards implementation and operation.

## Design stage study (PHS study 3)

The next safety study will be a full **HAZOP (PHA3).** This uses the information listed above, supplemented by details of the control system, key safety trips, draft batch sheets and operating instructions and the pre-HAZOP P&IDs. This is a major study involving a team of 6 or 7 led by an experienced member of the SHE team. Care will be taken in the selection of the stages of the operation to be subjected to a full HAZOP examination in order to optimise benefits from the man-hours used. Some commonplace operations will not be considered in detail but covered by comparison with similar steps, which have previously been the subject of a full study. Nevertheless, the study is expected to occupy the team for between 8 and 10 half-day sessions. A part of this study has been written up in Appendix A of the EPSC monograph (see Chapter 4) 'HAZOP: Guide to Best Practice'. This study will generate considerable follow-up work for the various groups. For example, the HAZOP study of one major stage of the process resulted in 34 actions; typically examination of the whole process (10 stages) may generate 150 actions. **Layer of Protection Analysis** is now used to confirm the preliminary SILs. The analysis is applied in each instance where a safety interlock has been identified as a safeguard either prior to or during the HAZOP study. The LOPA study will either confirm that the SIS in place is appropriate or indicate the PFD required (from the range of PFDs of 10%, 1% or 0.1%). As soon as the HAZOP questions are resolved a frozen, approved-for-construction set of P&IDS will be issued and the engineering work will be done. The **Relief and Blow down** study is finalised and confirmed, as is the **Process Hazard Assessment** documentation. Finally the Process Basis of Safety is written, showing how the final design meets the requirement of the CHA.

## Construction and precommissioning study (PHS 4 and 5)

A detailed **checklist based survey** of the plant will be done, based on the 'as-built P&IDs'; particular attention will be paid to new lines and to the isolation and removal of 'unwanted' lines. The list of actions from the **HAZOP** study must be confirmed as complete and signed off. Reviews will be prepared for the water and solvent flushes and for the cross-contamination checks that will be needed (which may require <10 ppm of specified products; the previous use of the plant greatly affects this). Water trials at temperatures of up to 50°C may be carried out as the construction is completed, but the list of actions from the **HAZOP** study must be confirmed as complete and signed off before solvent is allowed to enter the system. **Safety meetings** will be held with the operators and a training package developed. Before the plant is approved for active commissioning runs a **plant safety inspection** is required. This uses a basic **checklist** that has been tailored for this project.

## Post-start-up studies (PHS 6 and beyond)

It is possible that some modifications will be needed during the first campaign and more are likely in subsequent campaigns. These may result from early production problems that require a change in the batch size or from process optimisation. Modifications may be mechanical, process, operational or control. A **safety review** will be carried out for every change, using a **checklist** to decide what level of assessment and approval is needed. This may vary from a **simple risk assessment** made by a small team up to a repeat **HAZOP** study of a section of the modified plant or process. It is important to decide if the **basis of safety** or the **SIL** requirement has changed. A review of the changes proposed, at up to Operations Director level, is required before the changes can be implemented.

At regular intervals, for example every 5 years, the **HAZOP** study is revisited and consideration is given to the changing regulations and safety requirements imposed on the company to ensure that the process operates to a satisfactory standard.

Decommissioning and demolition issues do not arise within this project as it uses a multipurpose plant.

# Index

Note: Page numbers followed by *f* indicate figures and *t* indicate tables.

**H**

9780128195437